WILD FLOWERS OF MAJORCA, MINORCA AND IBIZA

Wild Flowers of Majorca, Minorca and Ibiza

With keys to the flora of the Balearic Islands

Written and illustrated by

ELSPETH BECKETT

A.A.BALKEMA / ROTTERDAM / BROOKFIELD / 1988

CIP-DATA KONINKLIJKE BIBLIOTHEEK, DEN HAAG

Beckett, Elspeth

Wild flowers of Majorca, Minorca and Ibiza: With keys to the flora of the Balearic Islands / Written and ill. by Elspeth Beckett. – Rotterdam [etc.]: Balkema. – Ill. With index, ref.
ISBN 90 6191 634 8 bound
SISO eu-span 585.2 UDC 582(210.7+460+262.1)
Subject headings: Majorca; flora / Minorca; flora / Ibiza; flora.

Published by
A.A.Balkema, P.O.Box 1675, 3000 BR Rotterdam, Netherlands
A.A.Balkema Publishers, Old Post Road, Brookfield, VT 05036, USA

ISBN 90 6191 634 8

Printed in the Netherlands

Contents

List of plates and figures

Preface

It would seem from the bibliography of main floras in the *Flora Europaea* that the Balearic Islands have had more than their share of botanical attention. Two out of fifty nine floras listed for the whole of Europe refer to the Balearic Islands only, and the islands are also included in a Flora of Spain. This is a lot for islands less than one three hundredth of the area of Europe.

Barceló's *Flora de las Islas Baleares*, however, published 1879-1881, has no illustrations, and Knoche's *Flora Balearica*, published 1921-1923, has no illustrations or descriptions. Knoche's list of Balearic species, like Barceló's, is difficult to use because many of the names of plants are obsolete and difficult to equate with modern names.

Since the publication of the *Flora Europaea* Bonafè's superb and complete *Flora de Mallorca* has been published. This is partly illustrated, and has complete descriptions of all the plants known to occur in Majorca. However it is very large, and written in Catalan, the language of the islands; a superb work for the specialist with a knowledge of the language, but not really suitable for others. nor very portable.

Bonner's excellent *Plants of the Balearic Islands* is useful for the identification of a few of the plants, but is more of an introduction to the plant associations of the islands than a taxonomic guide.

There seemed to be a need for a handy sized reference book for any interested holiday maker (among the 4,000,000 or so who visit these islands in a year) to be able to identify the plants.

I go to the Balearic Islands for my holidays too. I made a key to the flora from the keys in the *Flora Europaea* for my own use in the first place, gradually adapting it as I got to know the plants better (a key designed to cover the very large numbers of species in the whole of Europe is not necessarily the best key for a tiny fraction of them). I also added species not included in the *Flora Europaea*, but listed since its publication by the Belgian botanist, J. Duvigneaud, and the Danish botanist Alfred Hansen, and a very few I found myself in Majorca.

The result is offered here in the hope that plant hunting in these islands will give others as much happiness as it has given me.

Acknowledgements

My indebtedness to the authors and publishers of the *Flora Europaea* will be apparent to anyone who reads this book. Without this monumental work the task of chasing descriptions of minor species in the literature would have been very daunting. Polunin's *Flowers of Europe* and Polunin and Smythies' *Flowers of South-West Europe* were constant companions in the early stages, with abundant photos, drawings and descriptions of a large number of the species here. Clapham, Tutin and Warburg's *Flora of the British Isles*, where about one third of the plants occurring in the Balearic Islands are described, also regularly made the trip with me. This was the book from which I taught myself to use botanical keys, and a treasury of information of all sorts, including descriptions of plants which are much more at home in the Mediterranean than in the odd situations where they occasionally turn up in Britain. Bonner's *Plants of the Balearic Islands* was particularly useful for the descriptions of plant associations, a great help in finding species which I might otherwise have missed and for the very full and helpful bibliography. The lists published by J. Duvigneaud and Alfred Hansen considerably expanded earlier lists, as well as indicating the distribution of plants between the islands.

Among people who have helped I would particularly like to thank the following botanists: Dr Frank White of the Oxford University Dept of Botany, Mrs Mary Briggs of the BSBI, Mr J.R.Press of the European Herbarium of the British Museum of Natural History, and Dr Humphrey Bowen and Dr Stephen Jury of Reading University. All of them gave a lot of advice, and help with identification of specimens. I would like to thank too Dr S.M.Walters, who twice gave me permission to help myself to Balearic species in Cambridge University botanic garden to paint from, and Mr Clive King who sent me photos of a plant I failed to find there – and many other people who helped with particular problems and particular plants, personally and by post, including Mrs Lorna Ferguson, Dr Susan Eden, Dr Paul Cannon, and Professor Tutin (who identified a specimen and wrote a helpful letter *by return of post*, a remarkable kindness from a great botanist to an unknown amateur). Dr Irene Ridge took an early edition of the key off to Minorca twice, returning with many useful suggestions for alteration and improvement (as well as a lot of helpful photos and specimens). Two artists deserve special thanks too, Mrs Audrey Blackman and Miss Sarah Hosking. Both gave me much encouragement and advice at a stage when I was tempted to give up. Also heavily and sometimes uncomfortably involved were family and friends, who bore with periods when my interest in the Balearic flora would perhaps have been more appropriately described as an obsession; by their kindness and patience I got it out of my system and returned to a more normal range of interests.

Last and not least, my thanks are due to Mr A.T.Balkema, who saw fit to publish the result, and gave much valuable advice during the preparation.

How to use this book

Experienced botanists
This book is written by an amateur, and there are sure to be mistakes. The keys are based on information from all the books in the Bibliography, but the *Flora Europaea* has been the only source for many species.

You can of course go straight to the keys, but you will often find it quicker to use the illustrations as a short cut, or to use the index if you recognize the genus.

The keys include the plants listed for the Balearic Islands in:

1. *Flora Europaea*
2. Knoche's *Flora Balearica*
3. Bonafè's *Flora de Mallorca* (especially his Appendix II)
4. Duvigneaud's *Catalogue Provisoire de la Flore des Baléares*, 2nd edition (the 1st edition is the main source for Bonafè's Appendix II)
5. Hansen's additions and corrections to No. 4.

A few have been omitted because the records seemed to be very old or dubious, generally following Duvigneaud in this. I have also added a very few plants which I have found growing wild in Majorca; these have all been confirmed by experienced botanists. They are not necessarily new records, as I have not searched recent local and specialist publications.

Beginners
Before starting to identify plants it is essential to understand the way in which plants are classified.

Classification of plants
The official names of plants are in Latin, which was at one time conveniently international for botanists. Now fewer people study Latin, but it is at least equally inconvenient for all and therefore not liable to cause any ill-feeling about choice of language.

The use of an international name avoids problems with local names, which may be applied to several different plants, not because of ignorance, as country people distinguish between plants that are important to them with great skill, but because their main concern is not taxonomic botany. Plants tend among rural communities to be classified by uses. (Samphire is the English name for any succulent salty plant used as a pickle, and covers plants from three different families; Lousewort is any plant which seems to discourage lice, and Honeysuckle any plant whose flower is sucked for its sweetness).

All plants are individuals and there is a lot of variation between individual plants. However a group of plants which interbreed freely will be very much alike, and these will generally be recognized as plants of 'the same kind', which is what is meant by a species. Several plants which have many important botanical characteristics in common make up a genus. The genus and species names together make up the scientific name for a plant – *Rosa canina* is the dog rose, the generic name being *Rosa* (rose) and the specific name *canina* (canine).

Some of the Latin names are of great antiquity, and were used by the Romans or even their Sanskrit speaking forbears. Others are descriptive of botanical characteristics, or commemorate the name of a botanist (or his wife), or else may be a local name translated. There are no rules laid down for choosing the name of a plant, but there are rules strictly laid down for deciding which name has priority – this is normally the first name given from the time when Linnaeus first devised this system of naming (and it is surprising how many of them still retain the names he gave them). Sometimes, though, the naming has been complicated by reorganisation of plants into different genera as the result of more recent botanical research.

Until very recently the full name of a plant would have to include an abbreviation of the name of the botanist who first named it, and sometimes other botanists' names too if the original name of the plant had been modified. This was necessary to be sure that everybody understood in what sense the name was being used.

Supposing a plant, discovered by Mr Smith, was called *Bella rotundifolia* Smith, and a lot of botanists subsequently added specimens of this plant to herbaria. Then Mr Jones, a widely respected botanist, decided that the specimens actually included two distinct similar species, one with rather narrow leaves. The plants might then be split into two species, called *Bella rotundifolia* and *Bella angustifolia*. But there would be a problem because a lot of herbarium specimens of *B. angustifolia* would be already labelled *B. rotundifolia*.

So plants collected before the split would remain *B. rotundifolia* Smith, unless somebody took it upon himself to resort and reclassify all the old specimens. The plants collected after the split would be labelled *B. rotundifolia* (Smith) Jones and *B. angustifolia* Jones. In this way anybody examining a specimen labelled *B. rotundifolia* would know whether this name was used in the older, broader sense, or in the newer more restricted sense.

The whole business is actually even more complicated than this. The publication of the *Flora Europaea*, an enormous scholarly work involving the prolonged effort and co-operation of very many distinguished European botanists, with descriptions of practically every species of plant known to occur in Europe, has for the time being provided standard names for defined plants, though no doubt botanists are busily splitting many of these already.

All the Latin names here are used in the sense in which they are used in the *Flora Europaea*, except for a very few which are taken from F. Bonafè's *Flora de Mallorca* or other named sources. The names of these are given in inverted commas (for example '*Anthyllis fulgurans*'). Some of the names in inverted commas appear in the *Flora Europaea* too, though not as recognized species.

The classification of plants into larger groups than genera is not shown by the name. If you examine the flower of an apple tree and a bramble you will find that they have much in common with a rose. But there are important differences too, such as the development of an apple and a blackberry rather than a rose hip in the fruiting stage. These three plants all belong to the family Rosaceae. They are related, but not as closely as the plants in one genus. Families are grouped again into higher groupings. Within a short time of your beginning to take an interest in plants you will find yourself making reasonable guesses about these relationships which will considerably ease the business of identifying plants, and make plant hunting much more interesting.

Identification of plants

Now to identify your plant. Look at it carefully and see if you can find a fully open flower. Decide next whether this is radially symmetrical (like a wheel, any line drawn through the centre dividing it into two similar parts), or only symmetrical in one plane (like a face, the right side an approximate mirror image of the left). The first is called an *actinomorphic* flower and the second a *zygomorphic* flower. Then count the petals, and consider the colour before referring to the List of plates (see p.VI). Look out for flowers like daisies, thistles and dandelions, which are actually a whole lot of tiny flowers, or florets, set side by side on a receptacle, like the beads on a brooch.

If you are unable to decide which bit is the flower, try Plate 6. If there doesn't seem to be a flower, Plate 7 may help. If the plant seems to be cultivated as a crop or for ornament, try Plate 8.

If you are still unable to identify the plant (there are many not illustrated), you might like to try the keys. The keys do not usually include cultivated species, unless they are commonly found growing wild, but they do include practically all the wild species known to occur in the islands, other than recent additions to the list that have appeared only in specialized literature. On the page facing the illustrations you will find a reference to the key to the genus. If you have a flower that is very like one in the illustration, it might be worth checking whether this key will help. If not, try going back to the family key, which can always be recognized by the Latin termination -eae.

Many unfamiliar words will be found in the keys. These can be found in the glossary, and will soon seem familiar. The whole business is just a matter of practice.

How to use a key

A key is a series of mutually exclusive choices between two alternatives designed to differentiate between a given list of objects. In the system used here, the two alternatives may be separated on the page, but have the same number. After making a choice between the two numbers 1, the user chooses between the next pair of alternatives under the choice he has made. An example would be a key designed to differentiate between bird, mouse, man, dog and ant:

1. Animal with 6 limbs .... Ant
1. Animal with 4 limbs
  2. Usually stands on 2 hind legs only
    3. Fore limbs adapted for flying .... Bird
    3. Fore limbs not adapted for flying .... Man
  2. Usually stands on 4 limbs
    4. Length of body (excluding tail) less than 10cm .... Mouse
    4. Length of body (excluding tail) more than 10cm .... Dog

It should be noted that the key only differentiates between things it is designed to differentiate between. If the key above is used to identify a cat, it makes the cat a dog. Similarly if one of my keys is used to identify a plant which I have not considered it will lead to misidentification (or nowhere). Because of this, identification should be checked against a standard reference work if you think you have a new 'find' for the islands. Many will be found in Polunin's *Flowers of Europe* or Polunin and Smythies' *Flowers of South-West Europe*, but the standard authority is now the *Flora Europaea*, packed with keys and detailed descriptions.

Note to plant hunters

Wild plant species disappear continuously from all countries. A small part of this is due to climatic change, but as this also leads to gradual acclimatisation and establishment of marginal species, losses from this cause tend to be balanced by gains.

By far the largest losses in most countries are due to changes in land management, such as drainage of wetlands and destruction of woodland. In the Balearic Islands the widespread development of holiday resources invades or destroys many plant habitats: *Limonium majoricum* may have been completely exterminated by development – an endemic plant which grows nowhere else in the world. But even this is not all loss, and builders' rubble often provides a variety of microhabitats which make for rewarding plant-hunting.

In Majorca another factor has been even more destructive – the naturalization of huge numbers of goats, which will be met by anyone who walks in the mountains. Grazing animals, like fire, have helped to shape the distinctive flora of the islands. The balance has now been tipped in favour of the goats because they are no longer herded and culled for food. They are not choosy eaters, and only prickles and a chance seeding in crevices and inaccessible places save some of the rarer plants from extinction.

Some species can be reached by a botanist more easily than by a goat; the goats may be more nimble mountaineers, but even on hind legs they are not as tall as a man's outstretched arm, and they can not reach into cracks with fingers or use a walking-stick to hook a plant down.

Please leave what the goats have left. A quick sketch with a note of the colours, or a photograph, best with a close-up lens, may be as useful as a pressed specimen. When the plant is abundant a small specimen may be appropriate if you know it is abundant in many other sites too; otherwise the constant pilgrimage of botanists may threaten a plant with extinction.

The best plant hunters are keen conservationists.

Notes for key users

Abbreviations

annl	annual
BI	Balearic Islands
bnnl	biennial
FE	Flora Europaea
fl.	flower
fr.	fruit
lf, lvs, lflet	leaf, leaves, leaflet
Ma	Majorca
Mi	Minorca
I	Ibiza
pnnl	perennial
sp, ssp	species, subspecies
spklt	spikelet
usu.	usually
O	absent

Ma, Mi, and I refer to known occurrence in Majorca, Minorca and Ibiza; ?Ma means a dubious or very old record for Majorca. Occurrence in the smaller islands (Dragonera and Cabrera off the coast of Majorca, Formentera off the coast of Ibiza) is not mentioned here unless the smaller island is the only site in the Balearic Islands.

BI in FE means that the plant is listed in the *Flora Europaea* for these islands, but not in the other works in the Bibliography. The *Flora Europaea* does not usually differentiate between the islands.

White flowers
In the keys flowers are described as white if this is the usual or only colour. The key does not take account of white sports, which are common in many species. These will usually be found growing among more normally co-

loured flowers. White sports of *Geranium* species seem to be particularly common in Majorca.

Hybrids
These are often difficult to identify, and really a matter for the specialist. They are usually omitted here.

Subspecies
The keys are designed to identify species in the forms they occur here. Subspecies are not usually considered unless strikingly different or of special interest.

The flora of islands are of particular interest because geographical separation leads to the gradual evolution of new species. Varieties and subspecies are stages in the process. The primrose, *Primula vulgaris* is found in most of Europe; but the common ssp *vulgaris* does not occur in these islands, having evolved in Majorca into the distinctive ssp *balearica* with larger, paler and more fragrant flowers. The wild parsnip, *Pastinaca sativa*, another plant which is common throughout Europe, is absent from these islands too. Its niche is filled predominantly by another yellow umbellifer, fennel (*Foeniculum vulgare*), and the only local parsnip, *P. lucida*, is a plant growing mainly in the mountains. The foxglove *D. purpurea* is replaced too, by a smaller, very pretty hairy foxglove *D. dubia*.

Hybrid plants may occasionally become established too, and over the years compete effectively with the parents so that one or both are lost. The figwort *Scrophularia ramosissima*, according to the *Flora Europaea*, is found here only in a form transitional to *S. canina* ssp *canina*. The parent *S. canina* seems to be common.

Local names
These are arranged:- English; Spanish; German; French. The Spanish names include names used in the islands, Catalan, and Castilian names. As these often overlap, no attempt is made to separate them strictly, though the first names are more local, and the Castilian, if given, comes last.

Plants do not always have a popular name, especially in countries where they do not occur.

Measurements
Where only a single measurement is given for a part such as a leaf or a stem this refers to the length. When describing a flower it refers to the longest measurement (which may be either diameter or length) unless otherwise specified.

Keys to the flora of the Balearic Islands

Keys to plant families

1. Plants with an alternation of free-living generations, reproducing by spores (plants without flowers)........................... Pteridophyta p.9
1. Plants without an alternation of free-living generations, reproducing by seed (flowering plants)
 2. Ovules not enclosed in an ovary ('fir-trees')...... Gymnospermae p.10
 2. Ovules enclosed in ovary............................... Angiospermae p.10

MAIN KEY 1. PTERIDOPHYTA

Plants with alternation of free-living generations, reproducing by spores.

1. Stems jointed, circular sheath above each joint...... Equisetaceae p.24
1. Stems neither jointed nor sheathed (except sometimes by base of lf)
 2. Lvs not differentiated into blade and stalk
  3. Lvs forming basal rosette.................................. Isoetaceae p.23
  3. Lvs not forming basal rosette........................ Selaginellaceae p.23
 2. Lvs with distinct blade and stalk
  4. Lvs of 4 simple leaflets at top of long stalk .... Marsileaceae p.28
  4. Lvs not as above, usu. more or less triangular, much divided
   5. Fertile blades distinct from sterile leaves
    6. Fertile blade terminal on stem with sterile blade lateral .... Ophioglossaceae p.25
    6. Fertile and sterile blades separate .... Gymnogrammaceae p.25
   5. Fertile and sterile leaves similar (except sometimes in size)
    7. Sori covered by deflexed margin of leaf
     8. Lvs pinnate .... Pteridaceae p.25
     8. Lvs 2-3-pinnate
      9. Rhizome long; stalks solitary, erect below, with blade curving more or less horizontally at top .... Hypolepidaceae p.26
      9. Rhizome short; lvs arising in tufts, stalks not usu. erect
       10. Lflets more or less fan-shaped, widely spaced, with slender black stalks .... Adiantaceae p.25
       10. Lflets more or less circular or oval, not widely spaced ....

.... Sinopteridaceae p.25

7. Sori not covered by deflexed margin of lf
 11. Indusium absent
  12. Lvs with reddish-brown scales beneath...... Aspleniaceae p.26
  12. Lvs without scales on lower surface .... Polypodiaceae p.28
 11. Indusium present
  13. Indusium pointed egg-shaped and entire, or toothed (variously shaped) .... Athyriaceae p.27
  13. Indusium not as above
   14. Indusium circular with central boss, or kidney-shaped.......... .... Aspidiaceae p.27
   14. Indusium oval or linear .... Aspleniaceae p.26

MAIN KEY 2. GYMNOSPERMAE

Plants without an alternation of generations, reproducing by seed. Ovules exposed, not enclosed in an ovary (see fuller account p.29)

1. Lvs needle-like; fr. woody, globular or conical, with overlapping scales bearing ovules on upper surface ('fir-cone') .... Pinaceae p.29
1. Lvs not needle-like; fr. not as above
 2. Fr. bright red, not globular; solitary blackish seed visible at end
  3. Straggling, bushy plant, with no apparent lvs (lvs small, brown, scale-like) .... Ephedraceae p.31
  3. Small evergreen tree with linear, leathery lvs........... Taxaceae p.30
 2. Fr. globular red-brown to purplish (often with white or blue bloom), woody when ripe; seed borne on outside of fr....... Cupressaceae p.30

MAIN KEY 3. ANGIOSPERMAE

Ovules completely enclosed in an ovary

1. Aquatic plants, usu. growing with most of the plant submerged or floating.. p.11
1. Land plants; or plants rooted in water, lvs and infce mainly aerial
 2. Flowers without perianth, or with perianth calyx-like, with or without very minute or caducous petals
  3. Trees and shrubs... p.11
  3. Herbaceous plants.. p.12
 2. Flowers with obvious petals
  4. Corolla actinomorphic
   5. Corolla with 3 or 4 petals; or tubular with 3 or 4 lobes
    6. Corolla with 3 petals or lobes p.14

6. Corolla with 4 petals or lobes p.14
5. Corolla with 5 petals or more; or tubular with 5 lobes or more
7. Corolla with 5 petals or lobes p.15
7. Corolla with more than 5 petals or lobes p.20
4. Corolla zygomorphic p.21

Subkey 1. Aquatic plants, usually growing with most of the plant submerged or floating

1. Plant entirely free-floating, not rooted in mud
Floating lvs in long-stalked rosette .... Trapaceae p.96
2. Lvs not forming rosette, unstalked .... Lemnaceae p.192
1. Plant rooted in mud
3. Fls sessile or nearly so (except sometimes male fls)
4. Lvs with conspicuously spiny-toothed margins..... Najadaceae p.158
4. Lf margins not spiny-toothed, sometimes slightly toothed
5. Fr. with 2 spines at base when ripe............ Ceratophyllaceae p.47
5. Fr. not spiny even when ripe
6. Lvs all submerged .... Zannichelliaceae p.158
6. Terminal lvs forming floating rosette .... Callitrichaceae p.119
3. Infce with peduncle at least 1.5cm, usu. much more
7. Lvs all basal .... Alismataceae p.155
7. Lvs not all basal
8. Lvs whorled, segments pinnate, capillary .... Haloragaceae p.99
8. Lvs not as above
9. Fls solitary, axillary .... Elatinaceae p.96
9. Fls in terminal spike or branched infce
10. Fls in pedunculate axillary spikes....... Potamogetonaceae p.156
10. Fls in branched infce
11. Lvs up to 1mm wide .... Ruppiaceae p.157
11. Lvs 2-10mm wide (sometimes narrower in winter)
12. Rhizome covered in brush-like fibres..... Posidoniaceae p.157
12. Rhizome not covered in stiff fibres......... Zosteraceae p.157

Subkey 2. Trees and shrubs; flowers without a perianth, or with the perianth calyx-like, with or without very minute or caducous petals

1. Substantial trees
2. Lvs pinnate
3. Lvs with terminal lflet; fr. compressed, elongated, with wing at one end.. Oleaceae p.109
3. Lvs without terminal lflet; fr. large, elongated black pod................
.... Leguminosae p.66
2. Lvs simple

4. Lvs palmately lobed
   5. Trunk shedding bark in flakes .... Platanaceae p.63
   5. Trunk not shedding bark
      6. Lvs large, 10-20cm wide .... Moraceae p.33
      6. Lvs less than 5cm wide .... Aceraceae p.89
4. Lvs pinnately lobed or unlobed
   7. Lvs evergreen and leathery
      8. Lvs dark; fls one sexed, male with 6 stamens .... Fagaceae p.32
      8. Lvs pale bluish-green; fls hermaphrodite, stamens numerous .... Myrtaceae p.98
   7. Lvs deciduous, not leathery
      9. Leaves shallowly lobed.................................... Fagaceae p.32
      9. Leaves toothed, not lobed.............................. Ulmaceae p.32

1. Shrubs and woody climbers
   10. Lvs entire or toothed, or apparently absent
      11. Plant fleshy, mealy or glaucous, or any combination of these......... .... Chenopodiaceae p.37
      11. Plant not fleshy, mealy or glaucous
         12. Lvs shining; fr. dry with three horns .... Buxaceae p.90
         12. Lvs not or hardly shining; fr. globular, fleshy
            13. Fruit with single stone enclosing 2 seeds.......... Oleaceae p.109
            13. Fruit without stone (2-4 seeds).................... Rhamnaceae p.90
   10. Lvs deeply lobed or pinnate
      14. Woody climbing plant .... Vitaceae p.90
      14. Shrub
         15. Lvs fan-like, deeply palmately lobed .... Palmae p.191
         15. Lvs pinnate.. Anacardiaceae p.89

Subkey 3. Herbaceous plants; flowers without a perianth, or with the perianth calyx-like, with or without very minute or caducous petals

1. Fls very small, in axils of small chaffy bracts, usu. closely imbricate on a rhachis to form spklt; spklts themselves usu. variously aggregated into a compound infce; perianth absent or represented by minute scales or bristles, enlarging or not in fr.
   2. Spklts usu. subtended by two empty bracts, one above the other; stems usu. cylindrical, and hollow between nodes; lvs usu. with membranous ligule or fringe of hairs at junction between sheath and blade .... Gramineae p.168
   2. Spklts usu. with one bract; stems usu. solid often triquetrous; lvs without ligule .... Cyperaceae p.193
1. Fls not as above; perianth usu. present, or fls not in spklts
   3. Perianth actinomorphic, brownish or greenish, in 2 whorls of 3

4. Lvs mostly basal, linear, grass-like or cylindrical
 5. Fls in spikes; perianth fleshy .... Juncaginaceae p.156
 5. Fls not in spikes; perianth scarious .... Juncaceae p.165
4. Lvs not mostly basal, not linear
 6. Non-climbing plant; inner whorl or perianth enlarged and hardened in fr. .... Polygonaceae p.36
 6. Climbing plant; perianth segments not enlarging in fr.; fr. a red berry .... Dioscoriaceae p.163
3. Perianth not actinomorphic in 2 whorls of 3
 7. Fls arranged in dense globular heads or elongated spikes
  8. Fls in dense elongated spikes, sometimes interrupted
   9. Plant 1m or more
    10. Much branched plant with large oval lvs; fls greenish to pinkish .... Phytolaccaceae p.41
    10. Spike dense, brown, cylindrical; fls compressed .... Typhaceae p.193
   9. Plant less than 1m
    11. Fls densely arranged on club-like spadix, partially enclosed in sheath .... Araceae p.191
    11. Spike not club-like or sheathed .... Plantaginaceae p.133
  8. Fls in globular heads
   12. Male and female parts of infce separate (hermaphrodite fls often present too)
    13. Male and female fls in separate globular heads .... Sparganiaceae p.192
    13. Upper part head with female fls, lower male ..... Rosaceae p.64
   12. Fls all hermaphrodite .... Plantaginaceae p.133
 7. Fls not arranged in globular heads or elongated spikes
  14. Plants with milky latex; infce usu. umbellate .... Euphorbiaceae p.84
  14. Plants without milky latex; infce not umbellate
   15. Plant without chlorophyll .... Balanophoraceae p.35
   15. Plant with chlorophyll
    16. Plant very fleshy
     17. Stamens 20, connate in 5 bundles .... Aizooaceae p.41
     17. Stamens not connate in 5 bundles, usu. less than 20
      18. Stamens 1-5 .... Chenopodiaceae p.37
      18. Stamens 7-20.................................... Theligonaceae p.99
    16. Plant not or hardly fleshy
     19. Perianth of chaffy segments .... Amaranthaceae p.40
     19. Perianth not as above

20. Epicalyx of small lobes alternating with lobes of tubular calyx.. Lythraceae p.97
20. Epicalyx absent
21. Stamen 1 .... Rosaceae p.64
21. Stamens more than 1
22. At least some lvs in opposite pairs
23. Lvs with stinging hairs .... Urticaceae p.33
23. Lvs without stinging hairs
24. Plants dioecious, or with some fls unisexual .... Euphorbiaceae p.84
24. Fls hermaphrodite..................... Caryophyllaceae p.42
22. Lvs all alternate
25. Lvs often mealy; fr. dry, one-seeded, often enclosed in hardened perianth segments or 2 persistent bracteoles, enclosing female fls and become connate and hardened in fr. .... Chenopodiaceae p.37
25. Leaves not mealy; fr. not as above
26. Stamens 4.. Urticaceae p.33
26. Stamens 5 or more .... Euphorbiaceae p.84

Subkey 4. Corolla actinomorphic with 3 petals or lobes
1. Fls pink, white, or pale violet; plant entirely herbaceous; fr. of 1 or more dry carpels
2. Tiny annl; lvs 1-2mm, often reddish; fls axillary, sessile .... Crassulaceae p.62
2. Annl or pnnl; lvs 3cm or more, green; infce branched .... Alismataceae p.155
1. Fls yellow; twiggy shrubs with fleshy fr., eventually red
3. Fr. globular.. Santalaceae p.34
3. Fr. trilobed .... Cneoraceae p.88

Subkey 5. Corolla actinomorphic with 4 petals or lobes
1. Slender, straggling plants with lvs in whorls of 4 or more; fls usu. less than 4mm, in axillary racemes.............................. Rubiaceae p.111
1. Not as above
2. Parasitic plants without green lvs
3. Fls erect; stamens fused, anthers 8 .... Rafflesiaceae p.35
3. Fls nodding; stamens 8 .... Pyrolaceae p.105
2. Plants with green lvs, not parasitic
4. Stamens 2
5. Fls greenish-yellow or white............................ Oleaceae p.109
5. Fls deep pinkish-red...................................... Lythraceae p.97

4. Stamens more than 2
 6. Stamens 4
  7. Fls white to pinkish-white or deep pinkish-red
   8. Fls deep pinkish-red .... Lythraceae p.97
   8. Fls white to pinkish-white
    9. Small tree
     10. Lvs scale-like; fls in racemes .... Tamaricaceae p.95
     10. Lvs broad, spiny; fls in axillary cymes ...... Aquifoliaceae p.89
    9. Annl or pnnl herbs
     11. Plant fleshy .... Crassulaceae p.62
     11. Plant not fleshy
      12. Lvs opposite; fr. a capsule .... Linaceae p.83
      12. Lvs not opposite; fr. bivalved .... Cruciferae p.55
  7. Fls yellowish-white to brownish, or bright yellow
   13. Shrub .... Rhamnaceae p.90
   13. Herbaceous plant .... Plantaginaceae p.133
 6. Stamens more than 4
  14. Stamens 6 or 8
   15. Stamens 6, 2 with shorter filaments, between petals, 4 with longer filaments opposite petals .... Cruciferae p.55
   15. Stamens not arranged as above, filaments more or less equal
    16. Fls deep pink to pinkish – red or white (not purplish-yellow)
     17. Shrub; fls tubular .... Ericaceae p.105
     17. Herbaceous plant; petals free
      18. Petals notched, or broad, overlapping ...... Onagraceae p.98
      18. Petals entire, narrow, not overlapping...... Lythraceae p.97
    16. Fls yellow, greenish, or greenish- or purplish-yellow
     19. Shrubs or shrublets with simple lvs
      20. Lvs dark green, aromatic........................ Lauraceae p.51
      20. Lvs not dark green or aromatic .... Thymelaeaceae p.92
     19. Woody pnnls; lf segments narrow .... Rutaceae p.88
  14. Stamens more than 8
   21. Evergreen shrub, lvs dark, aromatic.............. Lauraceae p.51
   21. Not as above
    22. Fls white
     23. Lvs simple .... Capparidaceae p.53
     23. Lvs compound .... Ranunculaceae p.47
    22. Fls usu. red or yellow, not white .... Papaveraceae p.51

Subkey 6. Corolla actinomorphic with 5 petals or lobes

1. Infce of numerous small florets borne upon (usu. stalked) circular, conical, flat or concave base (receptacle), the whole surrounded by

calyx-like involucral bracts .... Compositae p.138

1. Infce not as above
  2. Fls any combination of green, yellow, yellowish or white
   3. Fls white
    4. Stamens 5 or less
     5. Petals free
      6. Fls in umbels, lvs usu. alternate, divided .... Umbelliferae p.99
      6. Fls not in umbels, lvs entire, usu. opposite, or apparently 0
       7. Plant not parasitic, lvs green .... Caryophyllaceae p.42
       7. Plant parasitic, lvs scale-like, not green .... Pyrolaceae p.105
     5. Petals united, at least below (tube sometimes very short)
      8. Fls pendant with reflexed petals
       9. Shrub with narrow, entire, leathery lvs ... Asclepiadaceae p.111
       9. Herbaceous plant
        10. Pnnl; lvs all basal .... Primulaceae p.105
        10. Annl; lvs not basal .... Solanaceae p.125
      8. Fls not pendant with reflexed petals
       11. Fls 4mm or less
        12. Perianth of 1 whorl .... Santalaceae p.34
        12. Perianth of 2 whorls, both corolla and calyx present
         13. Parasitic plant without chlorophyll; fls in globular clusters, stems thread-like twining .... Convolvulaceae p.114
         13. Lvs green; infce simple or branched, stems not thread-like
          14. Shrub .... Caprifoliaceae p.135
          14. Herbaceous plant .... Primulaceae p.105
       11. Fls 20mm or more
        15. Corolla funnel-shaped, hardly lobed ... Convolvulaceae p.114
        15. Corolla deeply lobed .... Apocynaceae p.110
    4. Stamens more than 5
     16. Trees or substantial shrubs (to 50cm or more)
      17. Shrubs with opposite lvs
       18. Lvs very shiny, smooth; petals and stamens creamish-white .... Myrtaceae p.98
       18. Lvs not or hardly shiny, indented with veins on upper surface; fls snow-white, stamens yellow to orange .... Cistaceae p.94
      17. Trees or shrubs with alternate lvs
       19. Lvs entire, shiny, evergreen; fls tubular .... Ericaceae p.105
       19. Lvs entire or lobed or pinnate, usu. not shiny and evergreen, if so, then pinnate .... Rosaceae p.64
     16. Herbaceous plants, or small, slightly woody pnnls
      20. Fr. capsule opening by teeth or valves..... Caryophyllaceae p.42
      20. Fr. of one or more separate carpels

21. Carpels numerous
22. Lvs palmate, lflets 5-7; petals pure white .... Rosaceae p.64
22. Lvs not palmate with 5-7 lflets; petals usu. yellow towards claw .... Ranunculaceae p.47
21. Carpels 4 or less
23. Lvs fleshy; carpels 4 .... Crassulaceae p.62
23. Lvs not or hardly fleshy; carpels 2, sometimes united .... Saxifragaceae p.63
3. Fls green, greenish-white, greenish-yellow, or yellow
24. Fls not pure yellow
25. Shrubs, not climbing
26. Lvs simple
27. Fr. 5-lobed .... Coriariaceae p.89
27. Fr. globular .... Rhamnaceae p.90
26. Lvs pinnate .... Anacardiaceae p.89
25. Herbaceous plants and climbing woody plants
28. Calyx and corolla present
29. Lvs all basal .... Primulaceae p.105
29. Lvs not all basal
30. Petals united at least at the base
31. Plant not climbing; corolla lobes much shorter than tube
32. Fls in erect terminal raceme .... Crassulaceae p.62
32. Fls in branched clusters .... Solanaceae p.125
31. Plant climbing or straggling; fls in few-fld axillary racemes; corolla lobes much longer than very short tube
33. Lvs unstalked, arranged in whorls .... Rubiaceae p.111
33. Lvs stalked, opposite .... Asclepiadaceae p.111
30. Petals free
34. Lvs fleshy; fls not in umbels .... Crassulaceae p.62
34. Lvs not fleshy; fls arranged in umbels
35. Plant climbing; lvs evergreen .... Araliaceae p.99
35. Plant not climbing; lvs not evergreen ... Umbelliferae p.99
28. Corolla absent; fls large; calyx petal-like, often with reddish margin .... Ranunculaceae p.47
24. Fls yellow
36. Petals united at least at base
37. Small tree or shrub; corolla tube much exceeding lobes .... Solanaceae p.125
37. Herbaceous plant; corolla tube not much exceeding lobes
38. Hairless plant .... Gentianaceae p.110
38. Plant hairy .... Scrophulariaceae p.127
36. Petals free

39. Fls in terminal spike-like raceme .... Rosaceae p.64
39. Fls not in terminal spike-like raceme
40. Stamens not more than 12
41. Stamens 5
42. Fls arranged in umbels........................ Umbelliferae p.99
42. Fls not arranged in umbels
43. Rough, fleshy procumbent plant .... Cucurbitaceae p.96
43. Erect plant, neither rough nor fleshy .... Linaceae p.83
41. Stamens 5-12
44. Lvs pinnate .... Zygophyllaceae p.83
44. Lvs not pinnate
45. Lvs entire, fleshy .... Portulacaceae p.42
45. Lvs trifoliate .... Oxalidaceae p.80
40. Stamens very numerous
46. Fr. of numerous free carpels
47. Lvs digitate with 5-7 lflets; fls axillary......... Rosaceae p.64
47. Lvs often palmately lobed, not as above; fls not axillary .... Ranunculaceae p.47
46. Fr. a capsule
48. Styles 3 or 5... Guttiferae p.93
48. Style 1 or 0.. Cistaceae p.94
2. Fls any combination of blue, pink, violet, or red; or white with markings in these colours
49. Main colour of fl. scarlet; or blue not or hardly tinged with purple
50. Main colour of fls scarlet
51. Shrub or small tree; fls 4cm across .... Punicaceae p.98
51. Herbaceous plant
52. Fls axillary... Primulaceae p.105
52. Fls not axillary................................... Ranunculaceae p.47
50. Main colour of fl. blue, not or hardly tinged with purple (sometimes slightly greenish)
53. Petals united into a tube at least at the base
54. Fls 2mm or less in flat-topped clusters....... Valerianaceae p.135
54. Fls exceeding 2mm
55. Fls in scorpioidal cymes .... Boraginaceae p.115
55. Fls axillary .... Primulaceae p.105
53. Petals free
56. Spiny maritime plant, fls in dense head .... Umbelliferae p.99
56. Not as above, plant not spiny
57. Lvs much divided .... Ranunculaceae p.47
57. Lvs entire .... Linaceae p.83
49. Main colour of fl. purplish or pinkish-blue, violet, pink; or white

with markings in these colours
58. Perianth of 1 whorl only (calyx petal-like, corolla 0)
59. Fls 5mm .... Primulaceae p.105
59. Fls 30mm or more.............................. Ranunculaceae p.47
58. Calyx and corolla present
60. Lvs all basal; fls small, numerous, in spikelets each with 1-5 fls subtended by 3 scale-like bracts, arranged more or less densely along branches of infce .... Limoniaceae p.107
60. Lvs usu. not all basal; infce not as above
61. Petals united, at least below
62. Tubular part of corolla as long as lobes or longer
63. Fls pink .... Convolvulaceae p.114
63. Fls violet or purplish-blue
64. Climbing plant............................ Convolvulaceae p.114
64. Not a climbing plant
65. Spiny shrub .... Solanaceae p.125
65. Annl .... Campanulaceae p.137
62. Tubular part of corolla shorter than lobes
66. Petals with one margin more strongly curved than the other ... . Apocynaceae p.110
66. Petals more or less symmetrically curved
67. Fls purplish-black; corona present..... Asclepiadaceae p.111
67. Fls not purplish-black; corona absent
68. Climbing pnnl; fruit a red berry............ Solanaceae p.125
68. Plant not climbing; fr. a capsule
69. Creeping pnnl; fls pink; capsule globose .. Primulaceae p.105
69. Erect annl; fls violet; capsule elongated....................... .... Campanulaceae p.137
61. Petals free
70. Stamens very numerous
71. Stamens united below into tube surrounding style .... Malvaceae p.91
71. Stamens free
72. Small tree
73. Fls tiny, in very dense cylindrical racemes...................... .... Tamaricaceae p.95
73. Fls 35-50mm in few-fld clusters .... Rosaceae p.64
72. Shrub
74. Spiny shrub .... Rosaceae p.64
74. Spineless shrubs, sometimes very small...... Cistaceae p.94
70. Stamens 10 or less
75. Plant fleshy.. Crassulaceae p.62

75. Plant not fleshy
76. Ovary with long beak, coiling or twisting spirally to release seeds when ripe .... Geraniaceae p.8
76. Ovary with short beak or 0, not coiling or twisting
77. Fls arranged in umbels
78. Lvs trifoliate, lflets entire .... Oxalidaceae p.80
78. Lvs various, not as above .... Umbelliferae p.99
77. Fls not arranged in umbels
79. Lvs trifoliate; stipules spiny .... Zygophyllaceae p.83
79. Lvs simple; stipules, if present, not spiny
80. Stamens 8-10.......................... Caryophyllaceae p.42
80. Stamens 4 or 6........................... Frankeniaceae p.96

Subkey 7. Corolla actinomorphic with more than 5 petals
1. Fls without sepals
2. Fls with two whorls of 3 brightly coloured petals, petals in one whorl often more or less differing from those in the other
3. Ovary inferior
4. Stamens 3; corolla without corona...................... Iridaceae p.164
4. Stamens 6; corolla often with corona.......... Amaryllidaceae p.163
3. Ovary superior
5. Woody plant with rosette of enormous bluish spiny leaves; stems up to 10m; infce terminal, fls yellowish............. Agavaceae p.162
5. Plant not usu. exceeding 2m, not woody; lvs various, not spiny; infce not as above .... Liliaceae p.158
2. Petals in one whorl of 6 or more, or corolla tubular
6. Lvs all basal, linear; corolla tubular with 6 small lobes .... Liliaceae p.158
6. Stem lvs present, much divided into linear segments; petals in one whorl .... Ranunculaceae p.47
1. Fls with sepals
7. Petals up to 12
8. Calyx tubular
9. Fls yellow; petals united at base .... Gentianaceae p.110
9. Fls purplish-pink; petals free .... Lythraceae p.97
8. Sepals free
10. Fls ca 10cm diam., pink; carpels 2 or 3............ Paeoniaceae p.51
10. Fls not exceeding 4cm, scarlet or yellow........ Ranunculaceae p.47
7. Petals very numerous
11. Plant woody, to 3m or more; fls borne on edges of large, glaucous, racket shaped joints; lvs reduced to prickles .... Cactaceae p.96
11. Plant not woody, up to 1m; lvs present, fleshy...... Aizoaceae p.41

Subkey 8. Corolla zygomorphic

1. Corolla with 2 whorls of 3 free lobes, the lower inner lobe much enlarged and usu. highly decorated; stamen 1, bearing 2 to 4 detachable masses of pollen (pollinia)................................ Orchidaceae p.196
1. Corolla with 5 segments or lobes or less; stamens usu. more than 1, pollen not aggregated into pollinia
2. Infce of numerous florets borne upon (usu. stalked) circular, conical, flat or concave base (receptacle), the whole surrounded by calyx-like involucral bracts
3. Florets often of 2 types differing in shape and colour as well as size; stamens 5, anthers fused together to form a tube round the style... Compositae p.138
3. Florets differing mainly in size, or not at all; stamens 4, free
4. Epicalyx present; corolla of 4 or 5 somewhat unequal lobes... Dipsacaceae p.136
4. Epicalyx absent; corolla strongly 2-lipped, the upper lip very short, the lower of 3 long lobes .... Globulariaceae p.131
2. Infce not involucrate
5. Petals free
6. Flower with one or more spurs
7. Stamens 8
8. Petals bright orange; calyx green............... Tropaeolaceae p.82
8. Perianth of two whorls, all blue, or some greenish... Ranunculaceae p.47
7. Stamens less than 8
9. Fls greenish-white or pinkish with darker markings; petals 4; stamens 2 .... Papaveraceae p.51
9. Fls violet; petals 5; stamens 5 .... Violaceae p.93
6. Flower not spurred
10. Zygomorphic fls only on outer margin of compound infce of which central fls are actinomorphic
11. Petals 4.. Cruciferae p.55
11. Petals 5... Umbelliferae p.99
10. Fls all zygomorphic
12. 'Pea flowers': 5 petals, upper usu. larger, held vertically; lower 2 projecting forwards and fused together by lower margins to form boat-shaped structure (keel), and 2 outer either appressed to keel or spreading.................... Leguminosae p.66
12. Not 'pea flowers'
13. Sepals 2, or 5 of which 2 are much enlarged
14. Two sepals much enlarged, bract-like, often coloured, more or less enclosing 3-petalled fl .... Polygalaceae p.88

14. Fls bright yellow, 4-petalled, not enclosed by sepals.. Papaveraceae p.51
13. Sepals 4-7, more or less equal
15. Fls white with orange spots, paired, axillary; stamens orange, much longer than petals .... Zygophyllaceae p.83
15. Fls white to pale yellowish in spike-like raceme; stamens whitish, hardly longer than petals .... Resedaceae p.61
5. Petals united at least at the base
16. Corolla long, funnel-shaped, usu. striped, with globular swelling at base, upper lip elongated.................... Aristolochiaceae p.35
16. Corolla tube various, with 4 or 5 lobes, often 2-lipped
17. Stamens 5
18. Fls arranged in scorpioidal cymes
19. Fls blue or purplish; fr. 4 one-seeded nutlets... Boraginaceae p.115
19. Fls pale yellowish-green; fr. many-seeded capsule.. Solanaceae p.125
18. Fls in terminal clusters or solitary
20. Tall climbing shrub .... Caprifoliaceae p.135
20. Small herbaceous or slightly woody plant
21. Plant much branched, with terminal clusters of deep blue to violet fls; calyx 10-lobed .... Primulaceae p.105
21. Fls solitary, terminal, pale blue or lilac; calyx 5-lobed.. Campanulaceae p.137
17. Stamens less than 5
22. Stamens 1 or 2
23. Fls pink to pale crimson in terminal clusters; stems usu. more or less dichotomously branched............ Valerianaceae p.135
23. Fls blue or white............................ Scrophulariaceae p.127
22. Stamens 4
24. Plants without green leaves................ Orobanchaceae p.131
24. Plants with green leaves
25. Fr. a capsule
26. Infce a robust spike with spiny bracts .... Acanthaceae p.131
26. Infce various, if a robust spike, then bracts not spiny
25. Fr. a drupe of 2-4 one-seeded nutlets
27. Style terminal
28. Upper lip of corolla very short, fls yellow ... Labiatae p.120
28. Corolla lips not very unequal; fls violet to pinkish... Verbenaceae p.119
27. Style gynobasic .... Labiatae p.120

Keys to genera of Pteridophyta

Each species of this subdivision of the plant kingdom is represented by two completely different free-living plants. One reproduces new individuals of the other, which in turn reproduces new individuals of the first. The smaller generation, the prothallus, is very small and easily overlooked.

In a typical fern a spore produced without fertilization gives rise to a small green prothallus, which is flat and heart-shaped and attached to the soil by root-like hairs or rhizoids. This produces male and female organs, the *antheridia* and *archegonia*, on the under surface. The antheridia each produce motile male gametes, and each archegonia one female gamete. Fertilisation starts the cell division which gives rise to the new fern, or sporophyte.

There are many variations. In the Gymnogrammaceae the prothallus is perennial and keeps producing new sporophytes, in others the prothallus disappears as the sporophyte develops. The prothallus may be entirely saprophytic and therefore not green (as in Ophioglossaceae), or may reproduce vegetatively as well as by fertilization.

SELAGINELLACEAE

Selaginella denticulata is the only species here, like a rather large-leaved moss, widespread in the mountains of Majorca, and thriving wherever moisture and shade occur together. (Ma, Mi, I)

SELAGINELLA Clubmoss; Pinchuita; Moosfarn; Selaginelle.

ISOETACEAE

These are tufted plants with hollow leaves with sporangia embedded in the base. A very ancient family, of considerable evolutionary interest, but not very striking to look at, and easily missed. There is only one genus here, *Isoetes*.

ISOETES Quillwort; Brachsenkraut in German.

Lf bases persisting as brown, papery scales .... *I. velata* (Mi)

Lf bases persisting as black, shiny scales, often 3-lobed, the lateral lobes spiny ....*I. durieui* (Ma, Mi)

EQUISETACEAE

These are perennial herbs with creeping rhizomes. The stems are jointed and longitudinally ridged, and have completely encircling sheaths, toothed at the upper end, above each joint. In many species there are also whorls of jointed branches similar to the stems but thinner at each joint. The yellow, pollen-like spores are borne under brown or blackish hexagonal scales, which fit together to make a cone at the top of either the branched stem, or a specialized thicker, yellowish-brown unbranched stem, or, as in the only common species here, at the tips of the branches. This family has only one genus, *Equisetum*.

EQUISETUM Horsetail; Asprella, Coa de cavall; Schachtehalm; Aprêle.

A. Key to cone-bearing fertile stems

1. Stems yellowish to brownish, not green
2. Sheaths numerous, with 20-30 teeth; cone 4-8cm .... *E. telmateia*(Ma, Mi)
2. Sheaths 4-6, with 6-12 teeth; cone 1-4cm .... *E. arvense* (Ma)
1. Stems green
3. Grooves 8-20; teeth of sheaths with a persistent hair-like apex; central hollow of stem 1/2 to 2/3 diam.; cone pointed.. *E. ramosissimum* (Ma, Mi, I)
3. Grooves 10-30; teeth of sheaths without hair-like apex; central hollow 4/5 diam. stem; cone blunt............................. *E. fluviatale* (?I)

B. Key to sterile stems

1. Stems whitish, branches many, regular .... *E. telmateia* (Ma, Mi)
1. Stems green, branches irregular, often few
2. Teeth of sheaths with narrow scarious border; cones 6-12mm, pointed .... *E. ramosissimum* (Ma, Mi, I)
2. Teeth of sheaths without a border; cones 10-20mm, blunt.. *E. fluviatale* (?I)

OPHIOGLOSSACEAE

The only species here is *Ophioglossum lusitanicum*, a tiny plant, rarely more than 5cm high. The solitary stem has a terminal spike with 3-12 sporangia on each side, and one or two simple narrow blades. (Ma, Mi, I)

OPHIOGLOSSUM Adder's tongue;-; Langue de serpent; Natternfarne.

SINOPTERIDACEAE

Two small ferns, one densely covered with hair-like yellow scales, the other smelling strongly of coumarin (the smell of new-mown hay). The scaly one is *Cheilanthes catanensis* (Ma, Mi, I), and the fragrant one *C. fragrans*. (Ma, I)

ADIANTACEAE

There is only one European species, *Adiantum capillus-veneris*. This is rare in Northern Europe, but common here, mainly around springs. (Ma, Mi, I)

Adiantum capillus-veneris Maidenhair fern; Culantrillo, Arañuela, Cabellos de Venus; -; Adiante.

PTERIDACEAE

Again only one species, *Pteris vittata*. This is a quite distinctive and handsome fern. The narrow fronds are borne almost vertically, and are up to 60cm, with 10 or more pairs of simple linear pinnae. (Ma)

GYMNOGRAMMACEAE

Anogramma leptophylla is a small annual fern arising from a short rhizome, not tufted (i.e. fronds arising singly, not from one point). It may be up to 20cm, but is generally much less. The prothallus is perennial. (Ma, Mi, I)

HYPOLEPIDACEAE

Pteridium aquilinum is common throughout most of the world. At one time it was valued for bedding for people and animals, but it is also a serious agricultural pest, partly because a lot of wild animals refuse to eat it. (Ma, Mi)

Pteridium aquilinum Bracken; Folguera, Helecho hembra; Saumfarn; Fougère.

ASPLENIACEAE

In this family the stem tends to be scaly and dark, with elongated or oval sori, but the range of form is very large.

KEY

1. Leaf entire, with or without auricles or lobes at the base .... *Phyllitis*
1. Leaf 1-3-pinnate
    2. Back of leaf densely covered in orange-brown scales; indusium absent .... *Ceterach officinarum* (Ma, Mi, I)
    2. Back of leaf not densely scaly; indusium present, same shape as sorus .... *Asplenium*

Ceterach officinarum Rusty-back; Herba daurada, Doradilla; Schriftfarn; Doradille.

PHYLLITIS Hart's tongue; Herba de la melsa; Hirschzunge; Langue de cerf.

Lamina without auricles or lobes at the base........ *P. scolopendrium* (Ma)
Lamina with auricles or lobes at the base .... *P. sagittata* (Ma, Mi, I)

ASPLENIUM Spleenwort; Falzia negra, Adianto; Streifenfarn; Adiante.

1. Lvs 1-pinnate
    2. Stalk with dense glandular hairs .... *A. petrarchae* (Ma, Mi, I)
    2. Stalk hairless
        3. Rhachis with narrow brown wing......... *A. trichomanes* (Ma, Mi, I)
        3. Rhachis with broad green wing................... *A. marinum* (Ma, Mi)
1. Lvs 2-3-pinnate
    4. Stalk blackish throughout
        5. Longest pinnae about middle of blade

6. Petiole ca 1cm; lamina with scattered glands.. '*A. majoricum*' (Ma)
6. Petiole longer than 1cm; lamina eglandular ..... *A. billotii* (Ma, Mi)

5. Basal pinnae longest
7. Lf blade triangular with straight sides, segments with acute teeth .... *A. adiantum-nigrum* (Ma, Mi)
7. Lf blades with concave sides and long tapering point,segments with long acuminate teeth .... *A. onopteris* (Ma, Mi, I)

4. Stalk green except sometimes at base
8. Lowest pinnae much shorter than middle ones.... *A. fontanum* (Ma)
8. Lowest pinnae longest......................... *A. ruta-muraria* (Ma, Mi)

(Many hybrid plants occur in Majorca, especially near Sóller: '*A. majoricum*' is possibly a hybrid).

ATHYRIACEAE

These are rather delicate, thin-bladed ferns.

KEY

Pinnae up to 15 on each side; indusium entire, inflated .... *Cystopteris fragilis* (Ma)
Pinnae more than 15 on each side; indusium toothed, not inflated........... .... *Athyrium filix-femina* (Ma)

ASPIDIACEAE

There are two genera and three species here, all medium sized ferns.

KEY

Lf uniformly dull green; pinnae with acute teeth; indusium kidney- shaped .... *Dryopteris villarii* (Ma)
Lf dark green above, paler below; pinnae with spiny point; indusium shield-like, with central boss .... *Polystichum*

POLYSTICHUM

Leaves flaccid, usually drooping; vein on which sorus is borne not or hardly continued beyond it .... *P. setiferum* (Ma)
Leaves leathery, rigid; vein continuing well beyond sorus...................... .... *P. aculeatum* (Ma)

POLYPODIACEAE

There is only one species here, *Polypodium australe*. The leaves are pinnate (or pinnatifid towards the top), usually pale green, arising singly from a creeping rhizome. It is common on walls, trees, and rocks, and is very like *Polypodium vulgare* (which probably does not occur here). (Ma, Mi, I)

POLYPODIUM Polypody; Falguereta; Tüpfelfarn, Engelsüss.

MARSILEACEAE

One species only here, *Marsilea strigosa*, a very rare plant of wet places. The stems have four spreading leaflets at the top, and sporangia at the lower end. (Ma)

Keys to genera of Gymnospermae

The plants so far have been spore-bearing. The rest are seed-bearing plants, the Spermatophytae, in which reproduction is brought about by seeds produced by fertilization of a female by a male gamete.

The Spermatophytes are divided into two major groups, the Gymnospermae and the Angiospermae. The word Gymnospermae comes from two Greek words, γυμνός, meaning naked, and σπέρμα, meaning seed, as opposed to Angiospermae, from the word 'αγγεῖον meaning a vessel. So, in theory at least, the Gymnospermae have naked seeds, while those of the Angiospermae are enclosed in a vessel, in this case an ovary. This distinction, first made in 1690, is of great importance in the study of the evolution of plants, but not of much practical use in the field.

Taking the plants occuring here, it is obvious that the seed of *Ephedra* is enclosed in a vessel, although the vessel has no lid; in *Pinus* the carpels have become modified to enclose the seeds, so they are no longer naked. In the Cupressaceae the seeds are naked on the outside of the fruit, but then so apparently are those of a strawberry, the fruit of an angiosperm, until you discover that strawberry pips are not seeds, but achenes, each containing a seed.

PINACEAE

There are two species of the genus *Pinus* here, probably recognised by most people by a combination of needle-like leaves and 'fir-cones'.

PINUS

Tree very regular, umbrella-shaped; trunk erect, bare of branches except at top; cone 10-15cm, globular .... *P. pinea* (I, introduced Ma, Mi)

Tree rather variably shaped, often not erect, commonly branched from lower part of trunk; cone 5-12cm, conical .... *P. halepensis* (Ma, Mi, I)

P. pinea Stone pine, Umbrella pine; Pi ver, Pino piñonero; Schirmkiefer; Pin parasol, Pin pignon.
P. halepensis Aleppo pine; Pi, Pi bord, Pino carrasco;-; Pin d'Alep.

Ibiza and Formentera used to be known as the Pityusae, Pine Islands, and have probably always been covered with *P. halepensis*, perhaps *P. pinea* too. In Majorca *P. halepensis*, very common round the coast and in the hills, has probably spread since large areas of Holm oak were destroyed by charcoal burners. In the coastal areas it is often planted for shade.

CUPRESSACEAE

There are two genera of this family here, *Cupressus* and *Juniperus*. *Cupressus sempervirens* is a native of Greece widely introduced here and throughout the Mediterranean. It can be found here both as tall, narrow fastigiate trees, especially around graveyards and churches, and in the more natural bushier form.

KEY

Fr. berry-like, with fleshy coalescent scales .... *Juniperus*
Fr. a woody cone, with woody scales .... *C. sempervirens* (introduced Ma)

Cupressus sempervirens Cypress; Xiprer, Ciprés; Cypresse; Cyprès.

JUNIPERUS Juniper; Sabina, Enebro; Wacholder; Genèvrier.
Lvs in whorls of 3, linear, sharp pointed .... *J. oxycedrus* (Ma, Mi, I)
Lvs scale-like, overlapping, in 4 or 5 rows .... *J. phoenicea* (Ma, Mi, I)

TAXACEAE

There is only one genus here. *Taxus baccata* (Yew; Teix, Tejo; Eibe; If). The huge sombre tree of English churchyards becomes in Majorca hardly more than a stunted bush of high mountains. This often happens with species better adapted to cooler climates – in the Himalayas the Yew retreats even higher.

One mountain in Majorca where it grows takes its name from the local name of the tree – Teix.

It is recognized by the reddish-brown, peeling bark, and shoots with two lateral rows of linear leaves, dark green on the upper surface and paler below. The flowers are unisexual, the female flower being followed by a cup-shaped red fleshy fruit, with the black seed visible at the open end. (Ma)

EPHEDRACEAE

This family is represented here by one strange plant, *Ephedra fragilis*, a straggling shrub often growing through *Pistacia lentiscus* or other shrubs. The upper part is a mass of green, often drooping, jointed stems. These produce small yellow flowers in the spring and berry-like fruit in the autumn, arranged in clusters at the joints. The leaves are reduced to small brown scales. (Ma, Mi, I)

Ephedra fragilis Joint pine; Trompera, Raïm de mer, Colmadillo; Meerträubchen; Raisin de mer, Uvette.

Keys to genera of Dicotyledons (Angiospermae)

Angiospermae are plants with the seed completely enclosed within an ovary until ripe. They are divided into two classes, the Dicotyledons developing from two 'seed-leaves', already present in the embryo, and the Monocotyledons from one (see next chapter).

FAGACEAE

Quercus is the only Genus here. *Q. ilex* forms extensive woods in the hillier parts of Majorca and Minorca, but not in Ibiza (the Island of Pines). The other species are not common. The most characteristic feature of this family is the fruit (acorn), like an egg in an egg-cup.

QUERCUS Oak; Roble; Eiche, Chêne.

1. Lvs evergreen and leathery, entire or toothed, not lobed
  2. Mature lvs grey-felted beneath; cup of acorn with spineless scales
    3. Mid-vein of leaves undulate; bark corky .... *Q. suber* (Mi)
    3. Mid-vein of leaves straight; bark not corky
      4. Leaves oblong; fruit bitter .... *Q. ilex* (Ma, Mi)
      4. Leaves rounded or ovate; fruit sweet .. *Q. rotundifolia* (I, ?Ma, Mi)
  2. Mature lvs hairless beneath; acorn cup with spreading, spine-tipped scales .... *Q. coccifera* (Ma, I)
1. Lvs semi-evergreen, not leathery, with 5-12 pairs of shallow lobes........ .... *Q. faginea* (Ma, ?introduced)

Q. ilex Holm oak; Alzina, Carrasca, Encina; Stein-eiche; Yeuse.
Q. coccifera Kermes oak; Carrasquilla, Coscoja; -; Chêne-kermès.

ULMACEAE

Ulmus (Elm; Om, Olmo; Ulme; Orme) and *Celtis* are the two genera here, both introduced, though often appearing to be wild.

KEY

At least some teeth at leaf-margin with secondary teeth; fruit dry .... *Ulmus minor* (Ma, Mi, I, introduced)

Teeth at leaf-margin all simple; fruit fleshy .... *Celtis australis* (Ma, I, introduced)

MORACEAE

There are no native species, but *Ficus carica* (Fig; Higuera; Feigenbaum; Figurier) is planted as a crop, and often naturalized in stony places. The large, lobed leaves are very late to appear, and the tree may seem dead when other trees are already in leaf. (Ma, Mi, I, introduced)

URTICACEAE

The plants in this family range from dull to frankly unattractive; the flowers are small and greenish, often unisexual, and the leaves are simple, often covered in stinging hairs, which produce a characteristic skin rash called Urticaria (even when *Urtica* is not to blame).

KEY

1. Lvs opposite, with stinging hairs .... *Urtica*
1. Lvs alternate, without stinging hairs
   2. Lvs 6mm or less.. *Soleirolia soleirolii* (Ma)
   2. Lvs more than 6mm .... *Parietaria*

Soleirolia soleirolii is better known as Helxine. It is curious that it seems to thrive in the moist climate of Northern Europe, sometimes taking over small gardens, while remaining a rare plant in the only places where it is native, the islands of Majorca, Sardinia and Corsica.

URTICA Stinging-nettle; Ortiga; Brennessel; Ortie.

1. Female fls in globular heads .... *U. pilulifera* (Ma, Mi, I)
1. All fls in elongated racemes
   2. Stipules 2 at each node................................ *U. dubia* (Ma, Mi, I)
   2. Stipules 4 at each node
      3. Pnnl; lf stalk about equalling blade .... *U. atrovirens* (Ma)
      3. Annl; lf blade shorter than stalk, at least in lower part of plant .... *U. urens* (Ma, Mi, I)

One of the attractions of these islands is the absence of *U. dioica*, the commonest nettle in most of Europe. *U. pilulifera* is a handsome plant, and not very common, and *U. atrovirens* is a rare mountain plant.

PARIETARIA Pellitory-of-the-wall; Albahaquilla; Glaskraut; Pariétaire. These are plants of walls and waste places. They present some problems for botanists here, as intermediates between the species given here certainly occur, and possibly other species and intermediates too, not easily distinguished.

1. Perennial; bracteoles connate at the base .... *P. diffusa* (Ma, Mi, I)
1. Annual; bracteoles free
   2. Lowest lf stalks usu. equalling or longer than blade; bracteoles shorter than perianth in fr.............................. *P. mauritanica* (Mi)
   2. Lowest lf stalks usu. shorter than blade; bracteoles equalling or longer than perianth in fr............................ *P. lusitanica* (Ma, Mi)

SANTALACEAE

These are herbs or woody plants, often partly parasitic on other plants. The flowers are very small.

KEY

Herbs; perianth tubular, 5-lobed; fr. dry and green................. *Thesium*
Shrubs; perianth of 3 or 4 free segments; fr. fleshy, becoming red *Osyris*

OSYRIS -; Ginesta de bolletes vermelles, Guardalobo; Harnstrauch; Rouvet.
Bracts leafy, persistent; fr. 5-7mm .... *O. alba* (Ma, Mi)
Bracts papery, soon falling; Fr. 7-10mm .... *O. quadripartita* (I)

THESIUM Bastard toadflax; -; Leinblatt; Thésion.
Pnnl; stems robust and much branched; nut with reticulate veins ....
.... *T. divaricatum* (Ma)
Annl; stems slender, branched only at base; nut longitudinally veined ......
.... *T. humile* (Ma, Mi, I)

LORANTHACEAE

Viscum album (Mistletoe; Muérdago; Mistel; Gui) seems to have occurred in Majorca at one time. There are no recent records.

ARISTOLOCHIACEAE

This is a family mainly of climbing or creeping plants, with simple alternate leaves. There are no petals, but the calyx is variously striped and decorated, tubular in the species here.

KEY

Fls actinomorphic, terminal........................ *Asarum europaeum* (?Ma)
Fls zygomorphic, axillary .... *Aristolochia*

ARISTOLOCHIA Birthwort; Aristolòquia; Osterluzei; Aristoloce.

1. Fls 2 or more together in lf axils..................... *A. clematitis* (Ma, Mi)
1. Fls solitary
  2. Lvs without stalks, encircling stems .... *A. rotunda* (Mi)
  2. Lvs stalked
    3. Fls brownish-yellow, stripes brown, limb reddish .... *A. bianorii* (Ma, Mi, endemic)
    3. Fls brownish-green, limb brownish-purple .... *A. longa* (Ma, Mi, I)

A. bianorii is named after E.C. Bianor (1859-1920), who made a fine collection of Majorcan plants, still preserved, as well as discovering this plant.

RAFFLESIACEAE

These plants are parasitic on the roots of other plants. *Cytinus* is the only genus here; the two species are parasitic on species of *Cistus* and *Halimium*. The flowers are subtended by two brightly coloured bracts, and have a tubular perianth with 5 lobes. There are no green leaves.

CYTINUS Local names Chupamieles, Chupera, Doncella.

Perianth bright yellow; bracts deep yellow to orange;.......................... .... *C. hypocistis* (Ma, Mi, I)
Perianth white or pale pink; bracts deep pinkish-red ...... *C. ruber* (Ma, I)

BALANOPHORACEAE

Another family of parasitic plants without chlorophyll, represented here by one species only, *Cynomorium coccineum*, a thick, fleshy pinkish-red plant with thousands of tiny flowers with 1-5 perianth segments arranged on a thick spathe-like stem. (I)

POLYGONACEAE

This family has small actinomorphic flowers with either 3 perianth segments, or 6 in two whorls. The segments are usually green and calyx-like, though some species of Polygonum have pink petal-like segments. The base of the leaf stalks often form membranous sheaths round the stem – these are called *ochreae*, and their colour and shape are important points for identification. The seed is solitary, and the fruit is often enclosed by persistent, sometimes enlarged, perianth segments.

KEY

1. Outer perianth segments with stout spines ... *Emex spinosa* (Ma, Mi, I)
1. Outer perianth segments not spiny
  2. Perianth segments 6, inner 3 much larger than outer in fr....... *Rumex*
  2. Perianth segments 5 or 6, the outer larger
    3. Outer perianth segments winged or keeled in fr. .... *Bilderdykia convolvulus* (Ma, Mi)
    3. Outer perianth segments not winged or keeled in fr. ..... *Polygonum*

POLYGONUM (Knotgrass, Bistort; Corretjola, Corregüela, Sanguinaria; Knöterich; Renouée) is a genus of rather unspectacular little plants, most with clusters of tiny pink or white flowers in the leaf axils, often partly concealed by the ochreae. In a few species the flowers are aggregated into terminal spikes, where they become more noticeable.

1. Ochreae white or silvery, becoming torn and ragged; infce few-fld, axillary
  2. Pnnl; stems woody at base
    3. Ochreae in flowering part stem longer than internodes, in middle of stem at least half internodes
      4. Ochreae with 8-12 strong, branched veins .... *P. maritimum* (Ma, Mi, I)
      4. Ochreae with fewer, feint veins...................... *P. romanum* (Ma)
    3. Ochreae shorter than all the internodes (except in unexpanded young stems)
      5. Lvs more than 6mm wide .... *P. oxyspermum* (Ma)
      5. Lvs less than 6mm wide................................ *P. romanum* (Ma)
  2. Annl; scarcely woody at base

6. Upper bracts scarious and shorter than fls............ *P. patulum* (Ma)
6. Bracts all green and lf-like, longer than fls
  7. Nut shining .... *P. oxyspermum* (Ma)
  7. Nut dull or hardly shining
    8. Petals broad, overlapping .... *P. aviculare* (Ma, Mi, I)
    8. Petals narrow, not or scarcely overlapping.. *P. rurivagum* (I, ?Ma)
1. Ochreae brown, opaque, entire or fringed; fls in terminal spikes
  9. Spikes lax, each fl. clearly visible............... *P. salicifolium* (Ma, Mi)
  9. Spikes crowded, dense, some fls more or less concealed by others
    10. Peduncles glandular; fls usu. white .... *P. lapathifolium* (Ma)
    10. Peduncles without glands; fls pink .... *P. persicaria* (Ma, Mi)

RUMEX (Dock; Romàs, Romaza; Ampfer; Romice) has small, usually greenish flowers arranged in whorls along the branches of an inflorescence, which usually has many branches. In flower the plants look rather weedy, but they can look very fine when the perianth segments turn bright red in the fruiting stage. Acid-tasting species are used as a pot-herb (Sorrel; Acedera; Sauerampfer; Oseille)

1. Lvs hastate or sagittate and tasting acid......... *R. intermedius* (Ma, Mi)
1. Lvs cordate, rounded or wedge-shaped at base; not or scarcely acid
  2. Valves more or less entire
    3. Valves 3-5mm,inverted heart-shaped; tubercles much narrower than valves; most whorls of fr. stem crowded.. *R. crispus* (Ma, Mi, I)
    3. Valves 2.5-3mm, elongated egg-shaped; tubercles often almost covering valve; all whorls of fr. stem remote .... *R. conglomeratus* (Ma, Mi, I)
  2. Valves with distinct teeth more than 1 mm long
    4. Pnnl; basal lvs large (up to 25cm) .... *R. obtusifolius* (Ma, I)
      5. Leaves thick, pedicels equalling valves....................... *R. pulcher*
      5. Leaves thin, longest pedicels 2x valves .... *R. obtusifolius*
    4. Annl; basal lvs very small .... *R. bucephalophorus* (Ma, Mi, I)

Rumex bucephalophorus has fruit like small whiskered red mice rather than heads of oxen. It grows in sandy places near the sea, including the higher parts of beaches.

CHENOPODIACEAE

The plants in this family include widespread agricultural weeds, and some important crops. There are also some highly specialized maritime plants, many with leaves reduced and fused in pairs to encircle the succulent stem, hardly recognizable as leaves at all.

KEY

1. Lvs flattened, thin bladed or slightly fleshy
 2. Fls mostly bisexual; fr. surrounded by 5 hardened perianth segments
  3. Some lvs toothed or lobed.................................... *Chenopodium*
  3. All lvs entire
   4. Lvs linear to narrow lanceolate............. *Bassia hyssopifolia* (Ma)
   4. Lvs very variable in shape, but broader, often rhomboidal...... *Beta*
 2. Fls unisexual; fr. enclosed in 2 enlarged, flattened bracteoles
  5. Annl plant
   6. Plant with male and female fls .... *Atriplex*
   6. Plant dioecious .... *Spinacia oleracea* (Ma, introduced)
  5. Shrub
   7. Fr. with 3-lobed bracteoles .... *Halimione portulacoides* (Ma, Mi, I)
   7. Fr. with rounded or kidney-shaped bracteoles................. *Atriplex*
1. Lvs thick and fleshy, or apparently absent if stems are fleshy
 8. Lvs apparent, alternate; stem not jointed
  9. Perianth fleshy in fr. .... *Suaeda*
  9. Perianth tough, not fleshy .... *Salsola*
 8. Lvs apparently absent; stem jointed
  10. Shrubby pnnl.. *Arthrocnemum*
  10. Annl.. *Salicornia*

BETA Beet; Acelga: Runkelrübe: Bette.

1. Receptacle in fr. hemispherical, with perianth segments 1-1.5mm .... *B. patellaris* (Ma)
1. Fr. receptacle bowl-shaped, perianth segments often exceeding 1.5mm
 2. Plant procumbent to ascending, up to 40cm; fr. perianth segments up to 5mm; infce of small 1-3 fld cymes, usu. developing only 1 fr. .... *B. macrocarpa* (Ma, Mi, I)
 2. Plant often erect, up to 200cm; fr. perianth segments up to 3cm; cymes 1-8 fld, usu. developing several fr. .... *B. vulgaris* (Ma, Mi, I)

CHENOPODIUM Goosefoot; Cenizo, Ceñiglo; Gänzefuss; Ansérine.

1. Plant pubescent, with sessile aromatic glands
 2. Calyx saccate, reticulate-veined....... *C. multifidum* (Ma, introduced)
 2. Calyx not saccate or reticulate-veined .... *C. ambrosioides* (Ma, Mi, I)
1. Plant hairless or mealy; not aromatic, not glandular

3. Seeds reddish-brown .... *C. glaucum* (Ma)
3. Seeds black
4. Plant foul-smelling when crushed.............. *C. vulvaria* (Ma, Mi, I)
4. Plant without distinctive smell
6. Lvs usu. much longer than broad; 'skin' of seed easily removed .... *C. album* (Ma, Mi)
6. Lvs usu. broader than long; 'skin' of seed only removable after boiling
7. Lvs usu. very mealy beneath; seeds shining... *C. opulifolium* (Ma)
7. Lvs very slightly mealy; seeds dull........... *C. murale* (Ma, Mi, I)

ATRIPLEX Orache; Armolls, Armuelles; Melde; Arroche.
1. Shrubby pnnl .... *A. halimus* (Ma, Mi, I)
1. Annl
2. Lower half bracteoles becoming hard in fr.; stems smooth or angled
3. Bracteoles reticulate-veined .... *A. tatarica* (Ma, Mi)
3. Bracteoles not reticulate-veined .... *A. rosea* (Ma, Mi, I)
2. Bracteoles not hardening in fr.; stem ridged
4. Lower lvs broadly wedge-shaped or truncated at base .... *A. hastata* (Ma, Mi, I)
4. Lower lvs narrow wedge-shaped at base, or tapering gradually to stalk .... *A. patula* (Ma, Mi, I)

ARTHROCNEMUM Local names, shared with *Salicornia*, are Glasswort, Samphire; Sosa, Polluelo, Alacranera; Glasschmalz, Queller; Salicorne.
1. Fls falling to leave tripartite hollow in segment; testa thin, hairy
2. Usu. green, becoming yellow or red; seeds with curved hairs .... *A. perenne* (Ma, Mi)
2. Usu. glaucous; seeds with conical hairs...... *A. fruticosum* (Ma, Mi, I)
1. Fls falling to leave undivided hollow in segment; testa hard, tuberculate .... *A. glaucum* (Ma)

SUAEDA Seablite; Sosa dura, Almajo; Sode; Soude.
1. Pnnl small shrub .... *S. vera* (Ma, Mi, I)
1. Annl
2. Perianth inflated in fr.; seed smooth .... *S. splendens* (Ma)
2. Perianth not inflated in fr.; seed with fine reticulate pattern .... *S. maritima* (Ma, Mi, I)

SALSOLA Saltwort; Salsora, Sosa; Salzkraut; Soude.
1. Annl

2. Plant prickly; lvs with spiny tip .... *S. kali* (Ma, Mi, I)
2. Plant not prickly; mucro at lf tip quite soft............ *S. soda* (Ma, Mi)
1. Shrubby pnnl
3. Lvs nearly all opposite; perianth segments blunt, fringed.................
.... *S. vermiculata* (Ma, Mi)
3. Lvs mostly alternate; perianth acute, not fringed ... *S. oppositifolia* (I)

SALICORNIA

S. europaea occurs (Ma, Mi, I), and probably other species too.

AMARANTHACEAE

The only genus here is *Amaranthus* (Amaranth; Amarant, Blet; Sammetblume, Fuchsschwanz; Amarante), annual or perennial herbs with small greenish flowers, and dry scarious petals. The flowers are inconspicuous individually, but in the species with a mainly terminal inflorescence they are often clustered in very large numbers, producing an appearance rather like a very dense, paniculate grass, such as a reed. The leaves, though, are not in the least grass-like, and are generally broad and entire with very prominent veins.

The plants are very popular in hot countries; although dull to look at, they remain green when drought has turned everything else yellow and brown. The name means 'not fading', and was applied by Pliny to a plant which was a symbol of immortality, probably a red-plumed species of this family, such as *A. caudatus* (Love-lies-bleeding). Milton may have had this plant in mind in a passage from *Paradise lost* – 'Immortal amarant, a flower which once in Paradise, fast by the Tree of Life, began to bloom, but, soon for man's offence to heaven removed ... with these, that never fade, the Spirits elect bind their resplendent locks'.

AMARANTHUS

1. Perianth segments 5
2. Infce a terminal panicle, apical part lfless
3. Plant usu. erect; fr. dehiscing transversely
4. Perianth segments blunt or truncate..
.... *A. retroflexus* (Ma, Mi, introduced)
4. Perianth segments acute....................... *A. cruentus* (Ma, Mi, I)
3. Plant usu. decumbent; fr. not dehiscent .... *A. muricatus* (Ma, Mi, I)
2. Infce of axillary clusters .... *A. blitoides* (Ma, I)
1. Perianth segments 2-3

5. Perianth half length of spine-tipped bracteoles .... *A. albus* (Ma, Mi, introduced)
5. Perianth longer than bracteoles; bracteoles not spiny
  6. Fr. dehiscing transversely .... *A. graecizans* (Ma, Mi, I)
  6. Fr. indehiscent or dehiscing irregularly
    7. Fr. smooth; seeds much smaller than fr. ..... *A. deflexus* (Ma, Mi, I)
    7. Fr. wrinkled; seeds almost as long as fr. .... *A. lividus* (Ma)

PHYTOLACCACEAE

The species here is *Phytolacca americana* (Pokeweed; Herba de l'oblea, Raïm de guilla; Kermesbeere; Raisin d'amérique). It was introduced from N. America for its berries, which yield a red dye used for dying food and clothes. The flowers are pink or green, in dense erect spikes arising in the angles between the branches of a tall leafy plant. (Ma, Mi, I, introduced)

AIZOACEAE

These are succulent plants with very fleshy leaves, and usually showy brightly coloured flowers. *Carpobrotus* (Hottentot Fig; Flor de cuchillo), a genus introduced from South Africa, is now a characteristic plant of the sea-shore here.

KEY

1. Petals absent; ovary superior .... *Aizoon hispanicum* (Ma, Mi, I)
1. Petals numerous; ovary inferior
  2. Fls 8cm or more across; fr. fleshy .... *Carpobrotus*
  2. Fls up to 3cm across
    3. Some leaves stalked
      4. Flowers purple............................ *Aptenia cordifolia* (Ma, Mi, I)
      4. Flowers white or pale yellow .... *Mesembryanthemum*
    3. Leaves all unstalked
      5. Leaves dark green; flowers pinkish..... *Disphyma crassifolium* (Mi)
      5. Leaves glaucous; flowers yellow......... *Lampranthus glaucus* (Ma)

CARPOBROTUS

Petals bright carmine-red.............. *C. acinaciformis* (Ma, I, introduced)
Petals pale pink, yellow or orange................ *C. edulis* (Ma, introduced)

MESEMBRYANTHEMUM

Whole plant covered in transparent swellings .... *M. crystallinum* (Ma, Mi, I, introduced)
Plant not covered in transparent swellings ..... *M. nodiflorum* (Ma, Mi, I)

Many related species are cultivated here, and can be found on tips etc.

PORTULACACEAE

Portulaca oleracea (Purslane; Verdolaga; Portulak; Pourpier) is a creeping, fleshy annual, with reddish stems, shining dark green leaves, and yellow flowers with 5 notched petals. A robust form, often more erect, is cultivated as a salad plant. (Ma, Mi, I)

CARYOPHYLLACEAE

This is a large family, with about twenty genera here, ranging from rather attractive flowers to small and weedy ones.

The characteristic features of the family are: opposite, linear leaves, often fused together in pairs at the base; dichotomous branching; five petals, pinkish-red to pink or white, often deeply lobed – in some species so deeply lobed that ten petals can easily be counted unless you look carefully.

The ovary is superior, and the fruit a dry capsule with a toothed opening at the top. The plants found here include many of these features, but not necessarily all, only the superior ovary being constant.

KEY

1. Stipules present
 2. All lvs alternate.................................. *Corrigiola telephiifolia* (Mi)
 2. At least some lvs opposite or in whorls
  3. Petals more than 2.5mm, entire, usu. pink or lilac .... *Spergularia*
  3. Petals less than 2mm (or if 2-3mm, then notched), white or greenish
   4. Sepals densely hairy all over...................................... *Herniaria*
   4. Sepals not densely hairy all over
    5. Sepals toothed or awned
     6. Sepals unawned, with a single bristly tooth on each side .... *Loeflingia hispanica* (I)
     6. Sepals awned.. *Paronychia*
    5. Sepals not toothed or awned

7. Bracts 6-10mm, exceeding and largely concealing fls .... *Paronychia*
7. Bracts much shorter, not or hardly concealing fls ..... *Polycarpon*
1. Stipules absent
8. Sepals free, or united only at base
9. Capsule teeth as many as styles
10. Styles fewer than sepals .... *Minuartia*
10. Styles as many as sepals .... *Sagina*
9. Capsule teeth twice as many as styles
11. Petals absent
12. Styles 3
13. Stalk with a single line of hairs down each internode ... *Stellaria*
13. Stalk uniformly hairy .... *Moehringia pentandra* (Ma, Mi)
12. Styles 5 .... *Cerastium*
11. Petals present
14. Petals entire or slightly notched
15. Petals rudimentary; seed with an oily appendage.................... .... *Moehringia pentandra* (Ma, Mi)
15. Petals not rudimentary; seed without an appendage
16. Stamens 5 .... *Cerastium*
16. Stamens 10... *Arenaria*
14. Petals bifid to at least 1/5 length
17. Lvs conspicuously hairy...................................... *Cerastium*
17. Lvs hairless or very slightly hairy .... *Stellaria*
8. Sepals united to form a distinct calyx tube
18. Calyx tube with veins at junction between sepals as well as mid-veins of sepals
19. Fls 3-5cm, reddish-purple; petals without coronal scales .... *Agrostemma githago* (Ma)
19. Fls less than 3cm; petals usu. with coronal scales .... *Silene*
18. Calyx tube without veins at junctions between sepals
20. Calyx tube with 5 wings .... *Vaccaria pyramidata* (Cabrera)
20. Calyx tube without wings
21. Calyx tube with papery junctions between sepals
22. Stems usu. unbranched; fls in terminal clusters; calyces concealed by brown papery bracts.......................... *Petrorhagia*
22. Stems branched above; calyces visible .... *Gypsophila pilosa* (?Ma)
21. Calyx tube smooth, tubular, without papery junctions
23. Epicalyx present .... *Dianthus rupicola* (Ma)
23. Epicalyx absent .... *Saponaria officinalis* (Ma, introduced)

ARENARIA Sandwort; Arenària; Sandkraut; Sabline.

1. Pnnl
  2. Lower lvs narrow lanceolate with long arista...... *A. grandiflora* (Ma)
  2. Lvs broadly egg-shaped or almost circular, not aristate .... *A. balearica* (Ma)
1. Annl or bnnl .... *A. serpyllifolia* (Ma, Mi)

MINUARTIA

1. Annl or bnnl; fls white
  2. Pedicels not longer than sepals; cymes crowded; seeds smooth .... *M. mediterranea* (Mi)
  2. Pedicels usu. longer than sepals; cymes lax; seeds slightly rough........ .... *M. hybrida* (Ma, Mi, I)
1. Pnnl; fls usu. pale pink .... *M. geniculata* (Ma)

STELLARIA Chickweed; Herba de canari; Sternmiere; Morgeli.

1. Sepals 5mm or more; stamens 10; seeds 1.3mm diam. or more .... *S. neglecta* (Ma)
1. Sepals 5mm or less; stamens 1-7; seeds 1.3mm diam. or less
  2. Sepals 4.5-5mm; petals usu. present; stamens with reddish-violet anthers.. *S. media* (Ma, Mi)
  2. Sepals 2-3.4mm; petals 0; stamens with greyish-violet anthers .... *S. pallida* (Ma, I)

CERASTIUM Mouse-ear chickweed; Morrons blancs; Hornkraut; Céraiste.

1. Petals as long as sepals or longer
  2. Pnnl; usu. without glands .... *C. fontanum* (Ma, Mi)
  2. Annl; usu. with glands at least above
    3. Capsule 7-10mm, curved, longer than pedicel............................. .... *C. glomeratum* (Ma, Mi, I)
    3. Capsule 6-7mm, very slightly curved, pedicel longer than capsule .... *C. pumilum* (Ma)
1. Petals shorter than sepals
  4. Upper bracts scarious apart from small central green part............... .... *C. semidecandrum* (Ma, Mi)
  4. Upper bracts not scarious
    5. Sepals thinly hairy to tip .... *C. brachypetalum* (Ma)
    5. Sepals with scarious non-hairy tip
      6. Pedicels shorter than sepals; fls clustered ...... *C. siculum* (Ma, Mi)
      6. Pedicels longer than sepals; fls in open dichasia ... *C. diffusum* (Ma)

SAGINA Pearlwort; -; Mastkraut; Sagine.
Lvs with terminal bristle .... *S. apetala* (Ma, Mi, I)
Lvs with very short point or 0.......................... *S. maritima* (Ma, Mi, I)

PARONYCHIA Arracades or Asperilla locally.
1. Pnnl; plant with conspicuous silvery bracts, largely concealing fls
 2. Fls in axillary clusters; calyx lobes hooded and awned .... *P. argentea* (Ma, Mi, I)
 2. Fls in terminal clusters; calyx lobes neither hooded nor awned.. *P. capitata* (Ma, Mi, I)
1. Annl; bracts inconspicuous; fls easily visible.......... *P. echinulata* (Mi)

HERNIARIA Rupture wort; Herba colica, Herba dela pedra; Bruchkraut; Turquette.
Pnnl with woody stock... *H. incana* (?Ma)
Annl, not woody... *H. hirsuta* (Ma, Mi, I)

POLYCARPON
1. Pnnl with woody stock; seeds tuberculate ... *P. polycarpoides* (Ma, Mi)
1. Annl (occasionally bnnl or pnnl) never woody; seeds not tuberculate
 2. Infce condensed (usu. coastal)............... *P. alsinifolium* (Ma, Mi, I)
 2. Infce lax (usu. inland)........................ *P. tetraphyllum* (Ma, Mi, I)

SPERGULARIA Sand-spurrey; Arenaria roja; Spärkling; Spergulaire.
1. Capsule 7-8.5mm (salt-marshes, robust pnnl) ...... *S. media* (Ma, Mi, I)
1. Capsule 6mm or less (annl, bnnl or pnnl)
 2. Lvs with terminal bristle
  3. Pnnl; stipules 6-10mm; fls violet; stamens 10 (coastal sands) .... *S. fimbriata* (I)
  3. Annl or bnnl; stipules less than 6mm; fls pink or whitish; stamens usu. less than 10 (dry sandy,gravelly or rocky places)
   4. Plant eglandular, or glandular only above; stipules silvery; fls 3-5mm (often inland).................................. *S. rubra* (Ma, Mi, I)
   4. Plant very glandular; stipules not silvery; fls 2mm (coastal) .... *S. bocconii* (Ma, I)
 2. Lvs without a bristle, sometimes mucronate
  5. Sepals blunt; capsule 4-6mm (drier parts salt-marshes) .... *S. marina* (Ma, Mi, I)
  5. Sepals acute to mucronate; capsule 2.5-3.5mm (maritime sands) .... *S. heldreichii* (Ma)

SILENE (Catchfly, Campion; -; Leimkraut; Silène)
Many species have individual local names, particularly *S. vulgaris* (Bladder campion; Colissos, Alcanducca; Schaummohn, Klatschnelke; Coulichon).

The genus includes some very attractive flowers, although none of them are very large. The petals, always 5, are contracted at the base into a narrow stalk, called the claw, which runs down inside the tubular calyx to its attachment below the ovary. There are often scales at the junction of the limb and claw. The calyx is usually attached to the pedicel some way below the ovary, so that the pedicel appears to be continued inside the calyx to the ovary. This part is called the carpophore, and the length of the carpophore is often important in distinguishing the species.

1. Pnnl, with leafy vegetative shoots
  2. Calyx inflated, with 20 veins........................ *S.vulgaris* (Ma, Mi, I)
  2. Calyx not inflated, with 10 veins
    3. Fls reddish-purple .... *S. hifacensis* (I)
    3. Fls white .... *S.mollissima* (Ma, Mi, I)
1. Annl, without leafy vegetative shoots
  4. Calyx white to reddish, with darker veins; petals deep rose-red.......... .... *S. rubella* (Ma, Mi, I)
  4. Calyx green, rarely red tinged; petals deep pink to white, or absent
    5. Fls all or mostly solitary, terminal on unbranched or sparingly branched infce
      6. Plant sticky with glandular hairs............................ *S. littorea* (I)
      6. Plant pubescent, not sticky or glandular........ *S. sericea* (Ma, Mi, I)
    5. Fls not mostly solitary or terminal
      7. Carpophore 2mm or less
        8. Plant hispid; stems simple, with fls in subcapitate to fastigiate infce .... *S. disticha* (Ma, Mi)
        8. Plant hairy or pubescent, not hispid; stems branched; fls not in fastigiate to subcapitate infce
          9. Capsule in fr. at least 2 1/2 x as long as wide
            10. Fls in raceme-like monochasial cymes; calyx 9-13mm, pubescent with ascending hairs .... *S. nocturna* (Ma, Mi)
            10. Infce repeatedly divaricately branched; calyx 6-8mm, densely covered with short, spreading glandular hairs .... *S. sedoides* (Ma, Mi)
          9. Capsule in fr. less than 2 1/2 x as long as wide (usu. less than 2 x as long as wide)
            11. Petals entire or notched, exserted .... *S. gallica* (Ma, Mi, I)
            11. Petals 0, or included in the calyx, or bifid ..... *S. apetala*(Ma, I)

7. Carpophore 2-10mm (exceeding 4mm in all except *S. cerastioides*)
12. Carpophore 9-10mm .... *S. pseudatocion* (Ma)
12. Carpophore not exceeding 6mm
13. Calyx strongly contracted below capsule in fr. (and thus closely conforming to junction between carpophore and capsule)
14. Calyx 8-11mm.......................... *S. cerastioides* (Ma, Mi, I)
14. Calyx 14-17mm .... *S. bellidifolia* (Ma, Mi)
13. Calyx weakly contracted or not contracted below capsule in fr. (and thus concealing junction of carpophore with capsule)
15. Seeds 2mm or more, back deeply grooved between 2 undulate wings .... *S. secundiflora* (Ma, Mi, I)
15. Seeds 0.9-1.1mm, back deeply grooved but not winged .... *S. nicaeensis* (Ma)

PETRORHAGIA Proliferous pink; Clavellina; Felsenelke; Oeillet.
These are very similar, like very small pinks clustered together in heads, surrounded by brownish papery bracts.

Lf sheaths at least twice as long as wide; seeds not more than 1.3mm, covered with cylindrical papillae.............................. *P. velutina* (?Ma)
Lf sheaths less than twice as long as wide; seeds at least 1.3mm, reticulate or covered in low tubercles .... *P. nanteuilii* (Ma)

CERATOPHYLLACEAE

Ceratophyllum demersum is the only species here. It is a submerged aquatic plant with minutely and irregularly toothed leaves arranged in whorls. There are tiny stalkless unisexual flowers at the nodes. A charophyte (*Nitella* sp.) with no true flowers and untoothed leaves, otherwise outside the scope of this book, can be mistaken for it, and seems common. (Ma, Mi)

RANUNCULACEAE

One of the most obvious constant features of this large family is the numerous stamens. In many species the ovary is made up of a large number of free carpels, arranged spirally on the receptacle.

The perianth may include any combination of sepals, petals and nectaries. In the key following ‘petal’ includes petal-like nectaries.

KEY

1. Carpels 2-5, each several-seeded, dehiscent
  2. Fls green, or reddish- or pinkish-green .... *Helleborus*
  2. Fls blue
    3. Fls spurred .... *Delphinium*
    3. Fls without spur .... *Nigella damascena* (Ma, Mi, I)
1. Carpels numerous, indehiscent, one-seeded
  4. Stem lvs in a whorl, or absent
    5. Stem lvs in a whorl, fls blue or red...... *Anemone coronaria* (Ma, Mi)
    5. Stem lvs absent; fls yellow
      6. Lvs hairless; receptacle elongated .... *Myosurus minimus* (Ma)
      6. Lvs bristly beneath; achenes in a globose head, receptacle short...... .... *Ranunculus*
  4. Stem lvs present, sometimes small, not whorled
    7. Woody climbing plants, petals 4 .... *Clematis*
    7. Herbaceous plants, petals usu. more than 4
      8. Petals secreting nectar, yellow or white (usu. less than 8).............. .... *Ranunculus*
      8. Petals not secreting nectar, red or yellow (usu. 8 petals or more)..... .... *Adonis*

HELLEBORUS Bear's foot; Pelònia, Eleborastro, Hierba llavera; Nieswurz; Pied de Griffin, Patte d'ours.

Lvs with 7-9 segments .... *H. foetidus* (Ma)
Lvs with 3 segments .... *H. lividus* (Ma, endemic)

DELPHINIUM Larkspur; Matapoll, Paparra; Rittersporn; Pied d'Alouette.

1. Limb of upper petal with lateral wings; seeds numerous.................... .... *D. halteratum* (I)
1. Limb of upper petal unwinged; seeds few
  2. Fls pale blue .... *D. pictum* (Ma)
  2. Fls deep blue .... *D. staphisagria* (Ma, Mi, I)

CLEMATIS These are usually (always here) woody climbing plants with 4-petalled flowers succeeded by heads of fluffy seeds.

C. cirrhosa is very common, particularly covering stone walls along the roadsides. *C. flammula*, also widespread, has a penetrating heavy scent and has been a popular garden plant because of this since Elizabethan times. It must be appreciated here too, since it seems to have many more local names than most plants.

1. Fls solitary, nodding .... *C. cirrhosa* (Ma, Mi, I)
1. Fls clustered, most erect
  2. Leaves once pinnate .... *C. vitalba* (Ma)
  2. Leaves bipinnate .... *C. flammula* (Ma, Mi, I)

C. flammula Maiden's bower; Matadent, Bugiot, Vidriella, Gessami bord, Badiella, Muermera; Waldrebe; Herbe aux gueux, Flammèle, Clématite brûlante.

ADONIS

Fls red .... *A. annua* (Ma, Mi)
Fls yellow .... *A. microcarpa* (Ma, Mi, I)

Adonis annua Pheasant's eye; Gota de sang, Adonis negro; Teufelsauge; Goutte de sang.

RANUNCULUS

There are about 20 species of *Ranunculus* here, of which 4 are white-petalled aquatic plants, and the rest yellow-petalled plants of waterside or dry ground. Some are distinctive enough to be identified at any stage of development, but for most a head of well developed achenes is necessary.

1. Fls white; plant aquatic (or on mud where water has dried)
  2. Petals rarely exceeding 5mm, soon falling; only finely divided, submerged lvs present .... *R. trichophyllus* (Ma)
  2. Petals usu. 6mm or more, persistent; broad bladed floating lvs often present as well as finely divided submerged lvs
    3. Achenes 40-100, hairless, usu. with a thin membranous wing on the keel when mature .... *R. baudotii* (?Ma)
    3. Achenes usu. less than 40; immature achenes pubescent, mature achenes wingless
      4. Petals less than 10mm; nectaries circular; fr. stalks usu. 2-5cm, rarely exceeding stalk of subtending lf............ *R. aquatilis* (?Ma)
      4. Petals usu. more than 10mm; nectaries pear-shaped; fr. stalks usu. more than 5cm, often longer than stalk of subtending lf............... .... *R. peltatus* (?Ma, ?Mi)
1. Land or waterside plants with yellow petals and no finely divided submerged lvs
  5. Cauline and basal lvs entire, crenate, saw edged or toothed, but not lobed
    6. Lvs all basal, bristly haired beneath................... *R. bullatus* (Ma)
    6. Stem lvs present, sometimes small; lvs not bristly beneath

7. Petals 5, pale yellow .... *R. ophioglossifolius* (Ma, Mi)
7. Petals 8-12, bright golden yellow .... *R. ficaria* (Ma, Mi)

5. At least some lvs lobed or more deeply divided
  8. Receptacle elongated in fr., head of achenes more or less cylindrical
    9. Beak of achene minute (almost invisible without lens) .... *R. sceleratus* (Ma, Mi)
    9. Beak of achene easily visible
      10. Beak of achene almost straight, flattened below .... *R. paludosus* (Ma, I)
      10. Beak of achene curved or hooked
        11. Beak of achene short, hooked .... *R. bulbosus* (Ma)
        11. Beak of achene about half diam. of basal part, curved .... *R. monspeliacus* (Ma)
  8. Receptacle not elongated in fr.; head of achenes more or less globose
    12. Mature achenes without spines or tubercles (may have bulbous-based hairs)
      13. Plant not more than 20cm high, basal lvs up to 2cm only (rare mountain plant) .... *R. weyleri* (Ma, endemic)
      13. Plant generally more than 20cm high; basal lvs more than 2cm
        14. Pedicels smooth .... *R. macrophyllus* (Ma, Mi, I)
        14. Pedicels furrowed
          15. Plant with stolons; sepals not reflexed .... *R. repens* (Ma)
          15. Plant without stolons; sepals strongly reflexed .... *R. bulbosus* (Ma)
    12. Mature achenes with spines or tubercles
      16. Achenes spiny
        17. Longest spines on border of achene......... *R. arvensis* (Ma, Mi)
        17. No spines on border of achene, only on face.......................... .... *R. muricatus* (Ma, Mi, I)
      16. Achenes tuberculate
        18. Tubercles confined to a circle just inside border of achene........ .... *R. sardous* (Ma, Mi, I)
        18. Tubercles scattered all over face of achene
          19. Tubercles hooked .... *R. parviflorus* (Ma, Mi, I)
          19. Tubercles not hooked
            20. Pedicel swollen and often recurved in fr. .... *R. chius* (Ma)
            20. Pedicel neither swollen nor recurved in fr. .... *R. trilobus* (Ma, Mi, I)

R. ficaria Lesser celandine; Gatassa, Morro de porro, Hierba de las almorranas; Scharbockskraut; Eclaircette.

Ranunculus (White-flowered aquatic forms) Water crowfoot; Ranúnculo acuático; Wasser-Hahnenfuss; Renoncule; (Yellow-flowered forms) Buttercup; Embrenyarelles, Pata de gallina; Hahnenfuss; Renoncule.

PAEONIACEAE

The only genus in this family is *Paeonia* (Paeony; Pelonis, Rosa de muntanya, Rosa de sarna, Rosa montesina; Pfingstrose; Pivoine). In Majorca and Minorca there is a fine endemic plant *Paeonia cambessedesii*, named after an important Spanish botanist, J. Cambessèdes (1799-1863). This is now rather rare, a plant of the mountains relished by wild goats. (Ma, Mi, endemic).

LAURACEAE

Laurus nobilis is the only species here (Bay; Llor, Laurel; Lorbeer; Laurier). It is a plant more characteristic of the Eastern Mediterranean, comparatively rare in Majorca and Minorca. Its Latin name honours its high symbolic status in ancient Greece and Rome, where the Laurel wreath was a reward for excellence in many fields (although not always made of *Laurus*). (Ma, Mi)

PAPAVERACEAE

Poppies are one of the glories of spring, and most people probably recognize the red ones. Slightly fewer might place the yellow seaside-poppy among them, with its long seed-pod and smoky-blue foliage, and to see the likeness between these and the flowers of *Fumaria* and *Hypecoum* requires rather careful examination – some botanists place them in a separate family. All of them are 4 petalled (at least here), a character which they share with Cruciferae, but there are only 2 sepals. The ovary is superior and the fruit a capsule, opening by pores or valves.

KEY

1. Fls actinomorphic
2. Capsule length less than 10 x width .... *Papaver*

2. Capsule length more than 10 x width
  3. Fls violet .... *Roemeria hybrida* (Ma, I)
  3. Fls yellow or red
    4. Fls solitary, yellow or red .... *Glaucium*
    4. Fls several together, yellow............ *Chelidonium majus* (Ma, Mi)
1. Fls zygomorphic
  5. Fls yellow .... *Hypecoum*
  5. Fls pinkish-purple, white or geenish, usu. with dark red decoration
    6. Upper petal spurred, stigma bifid .... *Fumaria*
    6. Upper petal not spurred; stigma trifid......... *Platycarpos spicata* (I)

PAPAVER Poppy; Rosella, Ababol, Amapola; Mohn; Pavot
1. Plant glaucous; upper lvs clasping stem ..... *P. somniferum* (Ma, Mi, I)
1. Plant not or hardly glaucous; lvs not clasping stem
  2. Capsule bristly
    3. Capsule longer than wide .... *P. argemone* (Ma, I)
    3. Capsule more or less globular .... *P. hybridum* (Ma, Mi, I)
  2. Capsule hairless
    4. Capsule less than twice as long as wide.......... *P. rhoeas* (Ma, Mi, I)
    4. Capsule at least twice as long as wide
      5. Anthers violet.. *P. dubium* (Ma, Mi)
      5. Anthers yellow or yellow-brown .... *P. pinnatifidum* (Ma, Mi, I)

GLAUCIUM
Fls red .... *G. corniculatum* (Ma)
Fls yellow.. *G. flavum* (Ma, Mi, I)

G. flavum Horned poppy; Rosella marina, Amapola amarilla; Gelber Hornmohn; Pavot cornu.

HYPECOUM
1. Peduncles ribbed
  2. Fls 1cm or less; middle lobe of outer petals larger than outer lobes...... .... *H. procumbens* (Ma)
  2. Fls 1-1.5cm; outer petals with 3 equal lobes .... *H. imberbe* (Ma)
1. Peduncles smooth .... *H. pendulum* (Is. Formentor)

FUMARIA Fumitory; Colomina, Herba de fum, Hierba del conejo; Erdrauch; Fumeterre.
Names in Latin, Spanish (Catalan and Castilian), German and French suggest the 'smokiness' of the plant. Not so English, where local names in-

clude Babe-in-the-cradle, Faminterry, God's fingers and thumbs, Wax dolls, Jam tarts, but no smoke. One of the problems with common names is that the same name may be given to many different plants. *F. parviflora* is the common smoky one, much less common in Britain than further South. In Britain the commonest is the green *F. officinalis*.

1. Corolla 8mm or less; lf segments narrow
  2. Lf segments with channelled lobes
    3. Plant glaucous .... *F. parviflora* (Ma, Mi, I)
    3. Plant not glaucous .... *F. densiflora* (Ma, I)
  2. Lf segments with flat lobes
    4. Lvs green; sepals 2-3.5mm .... *F. officinalis* (Ma, Mi)
    4. Lvs often glaucous; sepals 1mm .... *F. vaillantii* (I)
1. Corolla 9mm or more; lf segments broad lanceolate or wedge-shaped
  5. Fr. smooth when dry
    6. Infce with about 12 fls .... *F. muralis* (I, ?Mi)
    6. Infce with 15 or more fls
      7. Fls white with dark red tips, with or without reddish markings .... *F. capreolata* (Ma, Mi, I)
      7. Fls pink, wings and tip dark red .... *F. bastardii* (Ma, Mi)
  5. Fr. wrinkled when dry
    8. Pedicels strongly recurved in fr.......................... *F. flabellata* (Mi)
    8. Pedicels spreading or erect in fr.
      9. Sepals 2-3 x 1-2mm .... *F. bicolor* (Mi)
      9. Sepals 3-5 x 1-3mm
        10. Fr. only slightly wrinkled when dry .... *F. muralis* (?Mi, I)
        10. Fr. very wrinkled
          11. Corolla white to pinkish-white; apex of fr. with notched beak .... *F. agraria* (Ma, Mi, I)
          11. Corolla pink: apex of fruit almost unbeaked
            12. Upper petal blunt; fr. slightly ovoid .... *F. bella* (I)
            12. Upper petal slightly pointed; fr. rather angular .... *F. gaillardotii* (I)

CAPPARIDACEAE

Capparis (Caper; Taparera, Alcaparra; Kappernstrauch; Caprier) is the only genus here. It is found wild or cultivated (characteristically in rows between the trees in almond orchards) especially in the South East of Majorca and in Ibiza. The plant is a rather sprawling low shrub, covered in very lovely 4-petalled white flowers with masses of rosy stamens. The buds are pickled to produce capers.

PLATE 1. Actinomorphic pink and mauve or violet flowers with 5 petals.

	Key page
Caryophyllaceae	
1. *Spergularia rubra*	45
2. *Silene rubella*	46
3. *S. bellidifolia*	47
4. *S. sericea*	46
5. *S. gallica*	46
6. *S. cerastioides*	47
7. *S. secundiflora*	47
8. *Petrorhagia nantevilii*	47
Ranunculaceae	
9. *Helleborus lividus*	48
Crassulaceae	
10. *Sedum dasyphyllum*	63
Rosaceae	
11. *Rubus ulmifolius*	65
12. *Rosa rubiginosa*	65
13. *Prunus dulcis*	66
Geraniaceae	
14. *Geranium rotundifolium*	81
15. *G. purpureum*	82
16. *Erodium malacoides*	82
17. *E. reichardii*	82
18. *E. ciconium*	82
19. *E. cicutarium*	82
Zygophyllaceae	
20. *Fagonia cretica*	83
Malvaceae	
21. *Malva sylvestris*	91
22. *Lavatera cretica*	91
23. *L. arborea*	91
24. *L. maritima*	92
25. *L. trimestris*	92

	Key page
Cistaceae	
26. *Cistus albidus*	94
Tamaricaceae	
27. *Tamarix gallica*	96
Frankeniaceae	
28. *Frankenia hirsuta*	96
Primulaceae	
29. *Anagallis tenella*	106
Limoniaceae	
30. *Limonium duriusculum*	108
Gentianaceae	
31. *Centaurium erythraea*	110
32. *C. pulchellum*	110
Apocynaceae	
33. *Nerium oleander*	111
34. *Vinca difformis*	111
Asclepiadaceae	
35. *Vincetoxicum nigrum*	111
Convolvulaceae	
36. *Cuscuta epithymum*	114
37. *Convolvulus siculus*	115
38. *C. arvensis*	115
39. *C. althaeoides*	115
40. *Ipomoea purpurea*	115
Labiatae	
41. *Lavandula dentata*	125
Solanaceae	
42. *Solanum sodomeum*	127
Campanulaceae	
43. *Legousia hybrida*	138

28
41
9
39
5
2
1
8
7
29
6
4
43
40
37
3
38
36
32
31
10
30
35
33
11
34
26
42
27
12
24
16
25
23
13
18
17
15
14
22
21
19
20

CAPPARIS

Lvs more or less orbicular, mucro scarcely visible .... *C. spinosa* (Ma, I)
Lvs longer than wide, midrib prolonged into distinct mucro ....
.... *C. ovata* (Ma, I)

CRUCIFERAE

This family is familiar to most people at least in cultivation. It includes popular ornamental species such as *Cheiranthus cheirii* (Wallflower, Gilliflower; Violer groc, Alheli; Goldlack; Giroflée des murailles) and *Matthiola incana* (Stock, Gilliflower; Violer bord, Alheli blanco, Alheli vermell; Levkoje; Giroflée, Violer), and important vegetable and fodder crops; *Brassica oleracea* in numerous varieties (Cabbage, Cauliflower; Col de capell, Colflori; Weisskohl, Blumenkohl: Chou, Choufleur), *Brassica rapa* (Turnip, Rave, Ravenet; Rübenkohl; Rave, Navet) and *Raphanus sativus* (Radish, Rábano; Rettich; Radis).

Lobularia maritima (Sweet Alison, Alyssum; Xuclamel, Alis de mar, Murissa; Steinkraut; Corbeille d'argent, Blanquette), is cultivated for ornament in Northern Europe, and is a weed here. Some species are weeds anywhere, and you may be dismayed if you climb Massanella to find a major garden pest, *Cardamine hirsuta* at the peak, still showing off its very remarkable and effective method of seed dispersal.

The common features of the family can be seen in any of these: They are: racemose inflorescence (usually elongated and bractless), four petals (usually white, yellow or violet), usually six stamens, and a very characteristic fruit. The fruit is a capsule opening by two valves, often with a terminal beak. The beak may contain seeds and break up into transverse segments but usually most of the seeds are contained in the valvular part of the fruit, called a silicula when the length is less than three times the width, and a siliqua when the length is more than three times the width.

KEY

1. Tall, bristly annl; petals 1-2cm, white, yellow or violet, usu. with darker veins. Fr. cylindrical, 3-8-seeded, breaking into segments (and corresponding to the beak of most other species); small basal segment seedless .... *Raphanus raphanistrum* (Ma, Mi, I)
1. Not as above
 2. Petals white, or occasionally absent
  3. Fr. a silicula
   4. Fls irregular, 2 adjacent petals noticeably larger than the others;

fr. almost square .... *Iberis pinnata* (Ma)

4. All petals more or less equal
5. Septum of silicula across wider diameter
6. Petals deeply bifid .... *Erophila verna* (Ma)
6. Petals entire .... *Lobularia maritima* (Ma, Mi, I)
5. Septum of silicula across narrower diameter
7. Fr. a pair of reticulate, pitted hemispheres, with or without spines .... *Coronopus*
7. Fr. not as above
8. At least upper stem lvs auricled and clasping stem
9. Capsule winged.. *Thlaspi*
9. Capsule unwinged
10. Capsule obcordate (notch upwards).................... *Capsella*
10. Capsule cordate (notch downwards) .... *Cardaria draba* (Ma, Mi, I)
8. Upper stem lvs not auricled and clasping stem
11. Valves of fr. winged or strongly keeled
12. Plant hairless, or with few star-shaped hairs...................... .... *Hornungia petraea* (Ma)
12. Plant with scattered simple hairs
13. Petals as long as sepals or slightly longer........................ .... *Hymenolobus procumbens* (Ma, Mi, I)
13. Petals 11/2 – 2 x sepals................................... *Lepidium*
3. Fr. a siliqua
14. Lvs simple, with simple and/or branched hairs
15. Midrib of siliqua prominent; siliqua hairless............................ .... *Arabidopsis thaliana* (Ma, Mi)
15. Midrib of siliqua indistinct; siliqua hairy........................ *Arabis*
14. Lvs pinnatifid to pinnatisect, hairless, or with unbranched hairs only
16. Fls 15mm or more
17. Fr. 12-25mm, beak flattened, sabre-shaped, at least half length basal portion .... *Eruca*
17. Fr. 25-40mm, beak linear, very short (up to 4mm) .... *Diplotaxis*
16. Fls 6mm or less
18. Pnnl herb of running water ..... *Nasturtium officinale* (Ma, Mi, I)
18. Annl herb, not aquatic.............. *Cardamine hirsuta* (Ma, Mi, I)
2. Petals yellow, violet, lilac or pinkish
19. Petals yellow
20. Fr. with narrow 0-2-seeded basal portion below larger ridged and long-beaked globular portion ..... *Rapistrum rugosum* (Ma, Mi, I)

20. Not as above
21. Fr. almost spherical, spiny, with persistent style as long as spherical part .... *Succowia balearica* (Ma, Mi, I)
21. Not as above
22. Fr. a silicula
23. Fr. of flattened disc or discs
24. Fr. a single disc, pendant from pedicel .... *Clypeola jonthlaspi* (Ma, I)
24. Fr. 2 discs joined side-by-side, not pendant .... *Biscutella*
23. Fr. not disc-shaped
25. Fr. with broad, flat terminal appendage .... *Carrichtera annua* (Ma, Mi)
25. Fr. with persistent style or no terminal appendage .... *Neslia paniculata* (Ma, I)
22. Fr. a siliqua
26. Fls 25mm diam.
27. Fr. slightly flattened.. *Cheiranthus cheirii* (Ma, Mi,introduced)
27. Fr. 4-angled .... *Erysimum grandiflorum* (BI in FE)
26. Fls 20mm or less
28. Plant with finely pinnatisect lvs and star-shaped hairs .... *Descurainia sophia* (Ma)
28. Lvs various, not finely pinnatisect; hairs simple or absent
29. Seeds in one row in each cell
30. Valves 1-veined
31. Stem densely hairy; stem lvs not clasping stem .... *Erucastrum gallicum* (I)
31. Stem hairless, or if hairy, then leaves clasping stem
32. Plant completely hairless
33. Stem lvs few or absent; fls bright yellow.. *Brassica balearica* (Ma, endemic)
33. Stem lvs several, strongly clasping stem; fls very pale yellow .... *Conringia orientalis* (Ma)
32. Plant hairy at least below .... *Brassica*
30. Valves 3-7-veined, at least when young
34. Beak of fr. very short or absent .... *Sisymbrium*
34. Beak of fr. at least half length valves
35. Fr. erect and appressed to stem .... *Hirschfeldia incana* (Ma, Mi, I)
35. Fr. spreading.. *Sinapis*
29. Seeds in 2 rows in each cell

36. Beak broad and flat, at least half length valves .... *Eruca*
36. Beak cylindrical or conical, much shorter than valves (up to 4mm) .... *Diplotaxis*

19. Petals violet, lilac or pinkish
37. Lvs hairless, succulent; fls violet; fr. of 2 segments, upper conical with 2 broad teeth overlapping lower .. *Cakile maritima* (Ma, Mi, I)
37. Not as above
38. Fr. a silicula
39. Fr. almost square.................................... *Iberis pinnata* (Ma)
39. Fr. obcordate (notch at the top) .... *Capsella*
38. Fr. a siliqua
40. Plant completely hairless.............. *Moricandia arvensis* (Ma, I)
40. Plant hairy, at least below
41. Beak of fr. at least half length valves, flattened .... *Eruca*
41. Beak absent, but horns formed from thickened stigma lobes sometimes present
42. Star-shaped hairs present
43. Septum between valves completely opaque.......... *Malcolmia*
43. Margin of septum between valves translucent..................... .... *Maresia nana* (Ma, I)
42. Star-shaped hairs absent
44. Stigma in fr. thickened with knobs or horns .... *Matthiola*
44. Stigma in fr. not thickened .... *Malcolmia*

Raphanus raphanistrum White charlock, Runch; Citró, Erviana, Rabanillo, Jaramango menor, Ravenat; Hederich, Rettich; Ravenelle.
Cardaria draba Hoary pepperwort; Babol, Capellans, Pelitre; Pfeilkresse; Passerage drave.
Diplotaxis erucoides White wall rocket; Ravenissa blanca; Doppelsame.
Nasturtium officinale Watercress; Créixens, Glinxa; Brennenkresse; Cresson de fontaine.
Clypeola jonthlaspi Disc cress; Cabesa de mosca, Cabeza de mosquito; Schildkraut; Clypéole.
Descurainia sophia Flixweed; Sofia, Ajenjo loco, Sofia de los cirujanos; Besenrauke; Herbe de la sagesse.
Cakile maritima Sea rocket; Ruca de mar, Rucamar; Meersenf; Caquille.

SISYMBRIUM

1. Infce with bracts, at least in lower part
2. Fls usu. solitary; petals 2.5-3.5mm............ *S. runcinatum* (BI in FE)
2. Fls 1-3 together; petals 1.5-2.5mm .... *S. polyceratium* (Ma, I)

1. Infce without bracts
 3. Siliquae closely appressed to stem .... *S. officinale* (Ma, Mi, I)
 3. Siliquae spreading
  4. Lower lvs oval, entire or toothed .... *S. erysimoides* (Ma, I)
  4. Lower lvs deeply lobed or divided
   5. Siliquae 30-50mm, hairless .... *S. irio* (Ma, Mi, I)
   5. Siliquae 40-100mm, grey-green, hairy, becoming hairless .... *S. orientale* (Ma, Mi, I)

S. irio London rocket; Apagallums, Partesana, Rabanillo; Schlaffe; Vélaret.

S. officinale Hedge mustard; Herba dels cantors, Erissim; Weg-rauke; Vélar.

MALCOLMIA

Fr. 12-35mm .... *M. ramosissima* (Mi, I)
Fr. 35-80mm .... *M. maritima* (Mi)

MATTHIOLA

1. Fr. with 3 equal horns, at least 2mm .... *M. tricuspidata* (Ma, Mi)
1. Fr. without horns, or with only 2 horns more than 2mm
 2. Fr. with 2 horns up to 5mm, curved down at first, becoming horizontal .... *M. parviflora* (I)
 2. Not as above
  3. Siliqua compressed, oval in section
   4. Lower lvs deeply lobed; fr. with conspicuous glands .... *M. sinuata* (Ma, Mi, I)
   4. Lower lvs entire; fr. without glands .... *M. incana* (Ma, Mi, I, introduced)
  3. Siliqua circular in section .... *M. fruticulosa* (I)

ARABIS

1. Fls violet .... *A. verna* (Ma)
1. Fls white, (rarely pink)
 2. Lvs rounded at apex; fls 8-14 .... *A. muralis* (Ma)
 2. Lvs with an acute or obtuse point at apex; fls usu. more than 20 .... *A. hirsuta* (Ma, Mi)

CAPSELLA Shepherd's purse; Taleca de pastor, Pa i formatge, Sarronets de pastor, Paniquesillo; Hirtentaschelkraut; Bourse à pasteur.

Petals not or hardly longer than sepals, both usu. reddish .... *C. rubella* (Ma, Mi)

Petals usu. longer than sepals, white .... *C. bursa-pastoris* (Ma, Mi, I)

THLASPI Penny-cress; Traspic, Talaspico; Hellerkraut; Monnoyère, Tabouret.

Upper cauline lvs cordate and clasping stem at base; seeds smooth .... *Thlaspi perfoliatum* (Ma)

Upper cauline lvs arrow-shaped, clasping stem; seeds concentrically ridged .... *T. arvense* (?Ma)

BISCUTELLA Buckler mustard; Anteojero; Brillenschötchen; Lunetière.

1. Petals up to 15mm, spreading .... *B. auriculata* (Ma, I)
1. Petals not more than 8mm, erect
   2. Petals 4mm .... *B. frutescens* (I)
   2. Petals 5-7mm .... *B. sempervirens* (I)

LEPIDIUM Pepperwort; Morritort bord, Estiravelles; Kresse; Passerage.

1. Silicula 5mm or more, deeply notched and winged
   2. Sepals with white margin; petals 11/2x length sepals .... *L. spinosum* (Mi)
   2. Sepals without white margin; petals twice length sepals .... *L. sativum* (I)
1. Silicula 4mm or less, not or hardly winged and not notched .... *L. graminifolium* (Ma, Mi)

L. sativum is the cress of Mustard-and-cress, widely eaten in the seedling stage as a salad.

CORONOPUS Swine cress; Herba de sang; Krähenfuss; Coronope.

Fr. spiny and warty, not notched, in dense clusters opposite lvs .... *C. squamatus* (Ma)

Fr. notched and pitted, not spiny and warty, in elongated racemes opposite lvs .... *C. didymus* (Ma, Mi, I)

DIPLOTAXIS Wall rocket; Ruca pudenta; Doppelsame.

1. Petals white or pale violet, violet-veined ...... *D. erucoides* (Ma, Mi, I)
1. Petals yellow, sometimes with violet veins

2. Petals 8-15mm; fruiting pedicel about same length as fr. .... *D. tenuifolia* (Mi)
2. Petals 8mm or less; fruiting pedicel rarely more than half fr.
   3. Petals 5-8mm, about twice length sepals
      4. Lvs entire, toothed, or lobed...................... *D. muralis* (Ma, Mi)
      4. Lvs pinnatifid or pinnatisect....................... *D. catholica* (Ma, I)
   3. Petals hardly longer than sepals, up to 4mm *D. viminea* (Ma, Mi, I)

BRASSICA

1. All lvs glabrous
   2. Stem lvs absent or few; fls bright yellow... *B. balearica* (Ma, endemic)
   2. Stem lvs numerous; fls lemon yellow.... *B. oleracea* (Ma, introduced)
1. At least lower lvs with some hairs
   3. Upper stem lvs clasping stem
      4. Fls bright yellow, open fls overtopping buds.............................. .... *B. rapa* (Mi, introduced)
      4. Fls pale yellow or brownish, buds slightly overtopping open fls........ .... *B. napus* (I)
   3. Upper stem lvs not clasping stem .... *B. barrelieri* (I)

SINAPIS Mustard; Mostassa, Ravenissa groga; Senf; Moutarde.

Beak of fr. conical, straight, more than half length valves...................... .... *S. arvensis* (Ma, Mi, I)
Beak of fr. strongly compressed, sabre-like, equalling valves or longer...... .... *S. alba* (Ma)

ERUCA Rocket; Eruga, Ruca: Ruke: Roquette.

Beak of fr. about half length valves .... *E. sativa* (Ma, Mi, I)
Beak of fr. about equalling valves .... *E. vesicaria* (Ma, Mi, I)

RESEDACEAE

Reseda is the only genus here.

Perhaps the best known species is *R. odorata* (Mignonette; Gualda; Gabarrò; Réseda), a cottage garden plant grown for its scent. *R. luteola* (Dyer's Rocket, Weld; Cetro de Ceres, Galsa; Färber-Wau; Gaude) used to be cultivated for a yellow dye. It is now common in Europe, a rather distinctive plant with stiffly erect, usually unbranched, spike-like racemes of yellowish-green flowers.

KEY

1. Fls white, in dense conical racemes; sepals 5; capsule usu. with 4 apical lobes .... *R. alba* (Ma, Mi, I)
1. Fls white in lax racemes, or yellowish-green in denser racemes; sepals 4 or 6; capsule with 3 apical lobes
  2. Fls whitish, in short lax racemes; sepals 6; fr. drooping
    3. Lvs pale green, usu. unlobed, occasionally with one pair of lateral lobes .... *R. phyteuma* (Ma, Mi, I)
    3. Lvs dark green, many with several pairs of lobes .... *R. media* (BI in FE)
  2. Fls yellow or greenish-yellow; sepals 4 or 6; fr. erect
    4. Stems unbranched or sparingly branched; fls in dense, elongated stiffly erect racemes; sepals 4 .... *R. luteola* (Ma, Mi, I)
    4. Stem usu. much-branched; fls in short, compact, conical racemes; sepals 6 .... *R. lutea* (Ma, Mi, I)

CRASSULACEAE

Crassula and *Sedum* (Stonecrop; Herba de la roca, Siempreviva, Una de gato; Fettkraut; Orpin) are characteristic plants of this family – fleshy and drought resistant plants of walls, roofs and rocky places. *Umbilicus* (Wall pennywort; Llombrígol de Venus, Barretets; Venusnabel; Gobelets) is a distinctive genus with stiffly erect, many-flowered racemes of greenish-white tubular flowers, with at least some leaves with the stalk attached to the back rather than the edge, like an umbrella.

KEY

1. Petals and stamens in equal numbers
  2. Lvs opposite, connate; petals 3-4 .... *Crassula*
  2. Lvs not connate; petals 5 (rarely 4) .... *Sedum*
1. Stamens twice as many as petals
  3. Petals united to form a tube .... *Umbilicus*
  3. Petals free
    4. Plant woody at base; lvs forming terminal rosettes .... *Aeonium arboreum* (Ma, naturalized)
    4. Plant herbaceous; lvs not forming rosettes .... *Sedum*

CRASSULA

Fls sessile .... *C. tillaea* (Ma, Mi, I)
Pedicels present, longer than lvs .... *C. vaillantii* (Ma, Mi)

UMBILICUS

Fls drooping; stalk of basal lvs more or less centrally inserted .... *U. rupestris* (Ma, Mi)

Fls held horizontally; stalks of basal lvs usu. inserted towards one side .... *U. horizontalis* (Ma, Mi, I)

Other species or intermediates occur.

SEDUM

1. Lvs fleshy, cylindrical
  2. Stamens and petals in equal numbers
    3. Pnnl .... *S. album* (I)
    3. Annl
      4. Plant glandular-hairy .... *S. rubens* (Ma, Mi, I)
      4. Plant hairless .... *S. caespitosum* (?Ma, Mi)
  2. Stamens twice as many as petals
    5. Fls greenish-white to yellow .... *S. sediforme* (Ma, Mi, I)
    5. Fls pink or pinkish-white .... *S. dasyphyllum* (Ma, Mi, I)
1. Lvs fleshy, flattened .... *S. stellatum* (Ma)

SAXIFRAGACEAE

Saxifraga is the only genus here. The common species is rather small, and found almost throughout Europe, though an attractive little plant. It has fleshy, 3-lobed leaves.

KEY

Fls about 4mm diam.; bulbils absent.............. *S. tridactylites* (Ma, Mi, I)

Fls much larger; bulbils present at base of stem .... *S. corsica* (I)

PLATANACEAE

PLATANUS (Plane tree; Plátano; Platane; Platane) is the only genus. Both species here are introduced and sometimes naturalized. They are substantial trees which shed their bark in patches, so that the trunk is mottled in various shades of pale grey, brown and creamy white.

The fruit is a globular cluster of carpels dangling on a long stalk, usually two or more together.

KEY

Lvs lobed more than half way, wedge-shaped at the base; fruiting heads 3-6 .... *P. orientalis* (Ma)

Lvs lobed to half way at most, truncated or slightly notched at base; fruiting heads usu. 2 .... *P. hybrida* (ma)

ROSACEAE

This is a large and important family, including many plants cultivated for fruit – the apple (manzano; Apfel; pomme), peach (melocotón, prisco; Pfirsich; pêche), almond (ametller, almendro; Mandel; amande), strawberry (maduixera, fraulera, fresa; Erdbeere; fraise) and raspberry (frambueso; Himbeere; framboise), as well as the rose, probably one of the best loved and most widely cultivated plants of all, with a name that is much the same throughout Europe.

One of the features of the family is a large range of different fruits. The flowers are more uniform in pattern, generally with five petals, red, pink, white or yellow, and five sepals alternating with the petals. There are usually numerous stamens.

KEY

1. Trees or tall shrubs
 2. Thorns or prickles present, or both
  3. Prickly shrubs; stems often scrambling or arching
   4. Carpels free, fleshy when ripe, exposed over the surface of a convex receptacle .... *Rubus*
   4. Carpels enclosed in a fleshy receptacle .... *Rosa*
  3. Spiny shrubs or small trees
   5. Fls in corymbs (often up to 16 together) .... *Crataegus monogyna* (Ma, Mi)
   5. Fls solitary (occasionally 2 together) .... *Prunus*
 2. Thorns and prickles absent
  6. Carpels solitary; fr. with one-seeded stone .... *Prunus*
  6. Carpels 2-5; endocarp cartilaginous, not stony
   7. Fr. red, brownish-red or yellow .... *Sorbus*
   7. Fr. bluish-black .... *Amelanchier ovalis* (Ma)
1. Herbaceous plants, sometimes woody below
 8. Petals 0; sepals green or reddish-green
  9. Lvs palmately lobed .... *Aphanes*

9. Lvs pinnate................................ *Sanguisorba minor* (Ma, Mi, I)
8. Petals present, yellow or white
10. Fls yellow, in elongated spikes........ *Agrimonia eupatoria* (Ma, Mi)
10. Fls white and clustered, or bright yellow, solitary........... *Potentilla*

Crataegus monogyna Hawthorn; Espi blanc, Garguller, Majuelo; Weissdorn; Aubépine.
Amelanchier ovalis Snowy mespilus; Marful, Guillomera: Felsenmispel; Amélanchier.
Sanguisorba minor Salad burnet; Pimpinella, Herba balonera; Wiesenknopf; Pimprenelle.

RUBUS Bramble, Blackberry; Batzer, Bardissa; Brombeere; Ronce.
1. Drupelets 2-20, usu. with bluish-white bloom .... *R. caesius* (Ma, Mi, I)
1. Drupelets more than 20, bloom usu. absent
2. Lflets 3-5, pinnately arranged; petals usu. deep purplish-pink........... .... *R. ulmifolius* (Ma, Mi, I)
2. Lflets 5, digitately arranged, petals white or pink........................... .... *R. candicans* (?Mi)

ROSA (Exact identification of this genus is very difficult, and requires expert knowledge. This key will identify only typical specimens).
1. Styles connate in a column
2. Lvs leathery, evergreen; stylar column usu. hairy .... *R. sempervirens* (Ma, Mi, I)
2. Lvs not leathery, deciduous; stylar column hairless .... *R. arvensis* (Ma, Mi)
1. Styles free
3. Lflets slightly hairy, or hairless; glands, if present, confined to a few on teeth or on main veins on underside of lf
4. Lflets completely hairless
5. Pedicels without glands or hairs .... *R. canina* (?Ma)
5. Pedicels with stalked glands .... *R. pouzinii* (Ma)
4. Lflets slightly hairy............................ *R. corymbifera* (BI in FE)
3. Lflets very hairy or with numerous glands, especially beneath, sometimes both
6. Lflets with numerous apple-scented glands
7. Lflets rounded at base; pedicels rough with glandular hairs
8. Stems erect; styles hairy; petals deep reddish-pink; sepals persistent at least until fr. turns red .... *R. rubiginosa* (Ma)

8. Stems arching; styles hairless; petals pale pink; sepals soon falling .... *R. micrantha* (Ma)
7. Lflets wedge-shaped at base; pedicels without glandular hairs .... *R. agrestis* (Ma)
6. Lflets densely hairy, at least on the veins; glands resin-scented .... *R. scabruiscula* (?Ma)

(R. canina, R. corymbifera, R. micrantha, R. rubiginosa, R. scabriuscula are listed by Duvigneaud from the FE only, he gives no locality.)

POTENTILLA Cinquefoil, Tormentil; Peucrist, Pie de gallina; Fingerkraut; Quintrefeuille.
Fls white .... *P. caulescens* (Ma)
Fls yellow .... *P. reptans* (Ma, Mi, I)

APHANES Parsley piert; Peu-de-lléo, Pie de léon; Acker-Frauenmantel; Perce-pierre.
Stems slender; all lvs stalked .... *A. arvensis* (Ma)
Stems robust; at least upper lvs without stalks .... *A. floribunda* (Ma)

SORBUS
Lvs pinnate; fr. brownish-red to yellow, about 25mm .... *S. domestica* (Ma, I, cultivated and sometimes naturalized)
Lvs simple; fr. 8-15mm, scarlet .... *S. aria* (Ma)

Sorbus domestica Service tree; Ceridoler, Servera; Speierling; Sorbier.
S. aria Whitebeam; Muixera, Mostellar; Mehlbeere; Allier, Alisier.

PRUNUS
(This key does not include cultivated species, except *P. dulcis*, which does seed itself occasionally, and is such a major feature of the Majorcan landscape that it seemed a pity to leave it out).
Ripe fr. green, velvety.. *P. dulcis* (Ma, introduced, formerly the main crop)
Ripe fr. bluish-black, smooth .... *P. spinosa* (Ma, Mi)

LEGUMINOSAE

There are more species of this family in the Balearic Islands than of any other, even considering grasses.

It is divided into three large subfamilies. All except two species here,

Ceratonia siliqua and *Anagyris foetida*, belong to the same subfamily, which is described first.

The flower has 5 petals. The uppermost petal, the *standard*, is usually the largest, and sometimes distinctive in colour or decoration. It is generally held vertically. The two lowest petals are fused together by their lower edges to form a boat-like *keel*, in which lie the stamens, sometimes protruding at the end. The keel is usually directed forwards at right angles to the standard. The two lateral petals, or *wings*, are either appressed alongside the keel, or partly spreading, but directed forwards too.

The 10 stamens are fused below to form a tube. Sometimes one stamen remains free, outside the tube, which may be either oblique or truncated at the end, where the tips of the stamens become separate. The fruit is a pod. Peas (arvella, guisante; Erbse; pois) are Leguminosae. Sometimes the pod is different – it may be coiled up, or have constrictions along it, or be changed in various ways which make it more difficult to recognize, but the basic pea-pod, opening by two valves is the prototype for them all.

The leaves are variable. They are often pinnate, or with three leaflets. Purists make a distinction between *trifoliolate*, which means with 3 leaflets, and *trifoliate*, which means with three leaves; but trifoliate, used to mean 3 leaflets, is easier, and used here. Sometimes a leaf or a leaflet is modified to form a tendril, to enable the plant to cling on to surrounding plants and pull itself into the light, and sometimes the terminal leaflet becomes a spine to stop animals eating the plant; the spiny species here are the only ones really to thrive in the wilder, goat-infested parts of the mountains.

Stipules are always present, and sometimes much enlarged, so that they can be mistaken for a leaf.

Finally, the two oddities: *Anagyris foetida* has a flower much like the others,but the stamens are free, as are the 5 stamens of *Ceratonia siliqua*, which has no petals at all.

KEY

1. Large tree; corolla absent; stamens 5; pod about 20mm wide .... *Ceratonia siliqua* (Ma, introduced)
1. Small tree, shrub or herb; stamens 10; pod less than 20mm wide
  2. Lvs pinnate with even number of lflets, or with 2 lflets or none; terminal lflet or spine absent, terminal tendril sometimes present
    3. Two or more lflets present
      4. Tube of fused stamens transversely truncated; all stamens fused
        5. Stems smooth; wings of fl. fused to keel .... *Pisum sativum* (Ma, Mi, I, cultivated and escaping)

5. Stems winged or angled; wings of fl. free .... *Lathyrus*
4. Tube of fused stamens obliquely truncated; upper stamen partly free
6. Plant not climbing; seeds round, biconvex .... *Lens*
6. Plant climbing; seeds usu. oval or kidney-shaped .... *Vicia*
3. Leaflets absent .... *Lathyrus*
2. Lvs simple, or with terminal lflet or spine, or with lflets digitately arranged
7. Lvs with 5-11 lflets digitately arranged ...... *Lupinus micranthus* (Mi)
7. Lvs simple, pinnate, or with 1-3 lflets
8. All lvs simple or with 1 lflet (a fine distinction, unimportant here)
9. Herb .... *Scorpiurus*
9. Shrub
10. Spines present
11. Fls white with a violet keel.................................. *Anthyllis*
11. Fls yellow .... *Genista*
10. Spines absent
12. Fls 20-25mm; stems rush-like; lvs silky .... *Spartium junceum* (Mi, introduced Ma)
12. Fls much smaller; stems not rush-like; whole plant grey-silky ... . *Genista*
8. Some or all lvs with more than 1 lflet
13. Lvs with one lflet and lvs with 3 lflets both present
14. Lateral lflets much smaller than terminal
15. Fls on rigid vertical spikes with white woolly stems.... *Anthyllis*
15. Fls in long-pedunculate axillary clusters................ *Coronilla*
14. Lateral lflets not much smaller than terminal.............. *Ononis*
13. All lvs with 3 lflets or more
16. Lvs with 3 lflets; lateral lflets sometimes very small
17. Principal lateral veins ending at (often toothed) lf margin
18. Fls yellow, in sessile axillary clusters............................... .... *Trigonella monspeliaca* (Ma, I)
18. Fls variously coloured, not in sessile axillary clusters
19. Fls small, yellow, in elongated, lfless axillary racemes......... .... *Melilotus*
19. Fls not in elongated, lfless axillary racemes
20. Fls in dense terminal or axillary clusters, pedunculate or not
21. Pods coiled or strongly curved .... *Medicago*
21. Pods straight or slightly curved .... *Trifolium*
20. Fls variously arranged, not as above

22. Fls yellow, pods strongly coiled .... *Medicago*
22. Fls yellow, often with red veins, or pink and white .... *Ononis*

17. Principal lateral veins anastomosing, not reaching margin; margin entire
23. Plant spiny
24. Lvs of adult plant reduced to spine-tipped phyllodes (plant thus appearing lfless) .... *Ulex parviflorus* (I)
24. Lvs not as above, but often falling early
25. Toothed part of calyx pushed off by opening bud, leaving cup-like remnant (tip of bud surrounded by brownish calyx-teeth).. *Calicotome*
25. Toothed part of calyx not pushed off by opening bud
26. Calyx with 5 equal teeth, not 2-lipped .... *Anthyllis*
26. Calyx distinctly 2-lipped................................ *Genista*
23. Plant not spiny
27. Lflets distinctly spotted with glands, smelling of pitch; fls bluish-violet in dense, long-stalked clusters...................... .... *Psoralea bituminosa* (Ma, Mi, I)
27. Lvs not or hardly spotted with glands; fls yellow, white or red
28. Herbs, scrambling plants, or small shrubs
29. Fls white, with blackish keel .... *Dorycnium*
29. Fls yellow or red
30. Fls red, pod winged .... *Tetragonolobus*
30. Fls yellow, pod not winged
31. Keel much longer than standard .... *Genista*
31. Keel not longer than standard
32. Calyx in fr. much inflated .... *Anthyllis*
32. Calyx not inflated
33. Lvs with 5 leaflets; pod not densely hairy .... *Lotus*
33. Lvs with 3 leaflets; pod densely hairy .... *Argyrolobium zanonii* (Ma, I)
28. Substantial shrubs, or trees
34. Tall foetid shrub or small tree; stamens free; fr. a pod, 10-18cm long................... *Anagyris foetida* (Ma, Mi, I)
34. Shrub less than 3m; stamens variously connate; fr. not longer than 5cm
35. Upper lip of calyx deeply bilobed
36. Standard shorter than keel .... *Genista*

36. Standard longer than keel .... *Teline linifolia* (Mi, introduced Ma)
35. Upper lip of calyx with 2 short teeth .... *Chronanthus biflorus* (I)
16. At least some lvs pinnate (trifoliate lvs may be present too)
37. Fls yellow, yellow with bluish markings, or pale greenish-yellow
38. Fls yellow with bluish markings; fr. saw-like .... *Biserrula pelecinus* (Mi)
38. Fls yellow or pale greenish-yellow
39. Fls yellow
40. Pod breaking into segments when ripe
41. Segments of pod horseshoe-shaped .... *Hippocrepis*
41. Segments of pod not horseshoe-shaped
42. Fls in umbels of 10-20 .... *Coronilla*
42. Fls in clusters of 1-5 .... *Ornithopus*
40. Pod not breaking into segments when ripe .... *Lotus*
39. Fls pale greenish-yellow .... *Astragalus*
37. Fls pink, red, white or violet
43. Small cushion plant with terminal lflet usu. replaced by a spine; fls dirty pink, or white with red markings .... *Astragalus*
43. Not a cushion plant; spines absent
44. Fr. of 1-4 round flat discs, covered in spiny projections, discs breaking apart at maturity .... *Hedysarum*
44. Fr. not as above
45. Fr. inflated, with 1-2 large seeds, shaped like chickens trussed for the table .... *Cicer arietinum* (Ma, cultivated)
45. Fr. not as above
46. Fls in dense clusters, long pedunculate
47. Fl. heads deep pink, subtended by 2 leaf-like bracts; calyx inflated .... *Anthyllis*
47. Fls whitish to violet, without leafy bracts; calyx not inflated .... *Astragalus*
46. Fls 1-3 on short peduncle, bluish-violet to white ... *Lens*

Ceratonia siliqua Carob; Garrover, Algarrobera; Johannisbröt; Caroubier.

These beans may have been used once for weighing gold, as the word Carat seems to be derived from the name of the tree. The beans are a

valuable source of food for animals, and sometimes for people too. The German name follows the tradition that carob beans were the 'locusts' that kept St John the Baptist alive in the desert.
Anagyris foetida Bean trefoil; Garrover del dimoni, Colar de bruja; Stinkrauch; Bois puant.
Spartium junceum Spanish broom; Ginestera, Himestra; Spanischer Ginster; Genêt d'Espagne.
Psoralea bituminosa Pitch trefoil; Herba pudenta, Trébol bituminosa, Hierba cabruna; – ; Herbe au bitume.
Cicer arietinum Chick pea; Ciuronera, Garbazo; Zichererbse; Poischiche.
Anthyllis vulneraria Kidney vetch; Vulnerària; Wundklee; Vulnéraire. (Note: the only ssp of this plant here has pinkish-purple fls).

CALICOTOME

Fls mostly solitary; fr. almost hairless .... *C. spinosa* (Ma, Mi, I)
Fls in clusters of 2-15, or in bractless racemes; fr. hairy or silky, sometimes sparsely so .... *C. villosa* (Mi)

GENISTA Broom; Ginesta; Ginster; Genêt.

1. Plant spiny
  2. Branches spiny, axillary spines absent .... '*G. acanthoclada*' (Ma)
  2. Axillary spines present
    3. Lvs with spiny stipules; calyx silky............ *G. lucida* (Ma endemic)
    3. Stipules not spiny; calyx woolly .... *G. hirsuta* (I)
1. Plant without spines
  4. Lvs trifoliate; fr. about 10mm.............. *G. dorycnifolia* (I, endemic)
  4. Lvs simple; fr. usu. 20mm or more.. .... *G. cinerea* ssp *leptoclada* (Ma, endemic ssp)

ASTRAGALUS

1. Most lvs with rhachis ending in a spine .... *A. balearicus* (Ma, Mi, endemic)
1. Plant without spines
  2. Pnnl; fls white to purple .... *A. incanus* (?Mi)
  2. Annl; fls pale yellow or greenish-yellow
    3. Standard less than 4mm; pod egg-shaped............. *A. epiglottis* (?I)
    3. Standard more than 5mm; pod elongated, at least 4 x as long as wide
      4. Pod straight or slightly curved, more or less circular in section, grooved beneath .... *A. boeticus* (Ma, Mi, I)

4. Pod strongly curved, flattened, not grooved beneath .... *A. hamosus* (Ma, Mi, I)

VICIA Vetch; Vea, Veza; Wicke; Vesce.

1. Infce sessile or with peduncle much shorter than fls
2. Fls yellow
3. Standard densely hairy outside.......................... *V. hybrida* (Ma)
3. Standard hairless .... *V. lutea* (Ma, Mi)
2. Fls violet, purple or pinkish
4. Stipules half arrow-shaped, often with dark blotch .... *V. sativa* (Ma, Mi, I)
4. Stipules entire, not blotched,, *V. peregrina* (Ma, Mi, I)
1. Infce pedunculate, peduncle much longer than fls
5. Lflets 1-2 pairs; standard purple
6. Corolla 16-20mm, wings white .... *V. bithynica* (Ma, Mi)
6. Corolla 4-8mm, purple, white at base .... *V. bifoliata* (Ma, Mi endemic)
5. Lflets 3 pairs or more: fls various colours, standard usually same colour as wings or darker
7. Plant pnnl; corolla 10mm or more
8. Racemes 1-2-fld; corolla pale purple .... *V. monantha* (Ma, I)
8. Racemes with 4 fls or more
9. Fls reddish with darker tips; raceme shorter than subtending lf; fr. hairy .... *V. benghalensis* (Ma, Mi)
9. Fls violet or purple; raceme longer than subtending lf; fr. hairless .... *V. villosa* (Ma, Mi, I)
7. Plant annl; corolla up to 10mm (usu. less)
10. Most pods with 2 seeds
11. Pod about 10mm, pubescent...................... *V. hirsuta* (Ma, I)
11. Pod 12-20mm, hairless .... *V. disperma* (Mi)
10. Most pods with 3 seeds or more
12. Lvs with 5-10 pairs lflets; fls pale purple, white at base............. .... *V. leucantha* (Ma)
12. Lvs with 5 pairs lflets or less; fls pale blue to purple
13. 4 seeds in each pod
14. Racemes 1-2-fld; peduncles in fr. about same length as lvs; pod hairless .... *V. tetrasperma* (Ma, Mi)
14. Racemes with up to 6 fls; peduncle in fr. sometimes longer than lvs; pod usu. pubescent............ *V. pubescens* (Ma, Mi)
13. 5-8 seeds in each pod .... *V. tenuissima* (Ma, Mi, I)

LENS Lentil; Llentilla, Lentaja; Linse; Lentille.

Pod hairless; peduncle usu. aristate .... *L. nigricans* (I)
Pod pubescent; peduncle not aristate .... *L. ervoides* (BI in FE)

LATHYRUS Vetchling; Veça, Veza; Platterbse; Gesse.

1. Adult plant without lflets; tendrils arising from main stem; stipules lf-like .... *L. aphaca* (Ma, Mi, I)
1. Plant with lflets, at least in upper part; tendrils if present arising from end of lf; stipules much smaller than lvs
  2. Stem unwinged
    3. Lvs with up to 3 pairs lflets; fls pale blue or yellowish; pod 15-30mm .... *L. saxatilis* (Ma, I)
    3. Lvs with 1 pair lflets; fls not blue or yellowish; pod 30mm or more
      4. Peduncle 5-20mm; fls orange-red .... *L. sphaericus* (Ma, Mi)
      4. Peduncles 2-5mm; fls pale purple .... *L. inconspicuus* (?Ma)
  2. Stem winged, sometimes narrowly
    5. Fls yellow
      6. Lvs in lower part plant without lflets; dorsal margin of pod channelled or winged
        7. Dorsal margin of pod channelled .... *L. clymenum* (Ma, ?Mi)
        7. Dorsal margin of pod with 2 wings .... *L. ochrus* (Ma, Mi, I)
      6. Lvs with one pair lflets ; dorsal margin of pod not channelled or winged .... *L. annuus* (Ma, Mi)
    5. Fls not yellow
      8. Fls usu. 20-30mm in dense racemes of 5-15 bright carmine or pink fls .... *L. latifolius* (Ma)
      8. Fls less than 20mm, racemes with up to 5 fls; colour various
        9. Dorsal margin of pod channelled, winged or with 2 keels
          10. Fls red, reddish-purple, white, lilac or blue; lflets 1 (occasionally 2) pairs throughout
            11. Pod with 2 wings on each margin... *L. amphicarpos* (BI in FE)
            11. Pod with 2 keels, or 2 wings on dorsal margin only
              12. Fls brick red; fr. grooved on upper margin, unwinged .... *L. cicera* (Ma, Mi, I)
              12. Fls white, lilac or blue; fr. deeply grooved and 2-winged on upper margin .... *L. sativus* (I)
          10. Fls crimson with lilac or blue wings; lflets absent in lower part plant, 2-5 above .... *L. clymenum* (Ma, ?Mi)
        9. Dorsal margin of pod not channelled, winged or keeled
          13. Standard crimson with pink or white wings; lflets 2-4 pairs in upper part plant and none below ...... *L. articulatus* (Ma, Mi)

13. Fls orange-red; lflets 1 pair throughout plant .... *L. setifolius* (Ma)

ONONIS Restharrow; Ugons, Afrontallauradors, Gatuña; Hauheche; Bugrane.

1. Fls pink
2. Plant annl, without spines
3. Plant 15cm or less, solitary; pedicels strongly deflexed after flowering .... *O. reclinata* (Ma, Mi, I)
3. Plant 15cm or more, in dense terminal raceme; pedicels not deflexed after flowering
4. Corolla exceeding calyx .... *O. mitissima* (Ma, Mi)
4. Corolla equalling calyx or shorter............. *O. alopecuroides* (Ma)
2. Plant pnnl, sometimes spiny
5. Spines usu. present; stems erect or ascending, with one or two lines of hairs .... *O. spinosa* (Ma, Mi, I)
5. Plant rarely spiny, usu. procumbent; stem uniformly hairy .... *O. repens* (?Ma)
1. Fls yellow, often with red or purple veins on standard
6. Annl; stems herbaceous
7. Primary branches of infce aristate; pod up to 20mm
8. Pod curved, knobbly .... *O. ornithopodioides* (Ma, I)
8. Pod straight, sometimes slightly inflated, not knobbly.................. .... *O. viscosa* (Ma, Mi, I)
7. Primary branches of infce not aristate; pod 8-10mm .... *O. pubescens* (Ma, ?Mi)
6. Pnnl; stems woody at base
9. Corolla up to 12mm; pod 6-8mm
10. Fls sessile in lax spikes........................... *O. pusilla* (Ma, Mi, I)
10. Fls pedicellate in dense terminal raceme.................................. .... *O. minutissima* (Ma, Mi, I)
9. Corolla usu. more than 12mm; pod 10-25mm
11. Margins of lflets undulate; infce dense .... *O. crispa* (Ma, Mi)
11. Margins of lflets not undulate; infce lax...... *O. natrix* (Ma, I, ?Mi)

MELILOTUS

1. Fls 6mm or longer
2. Pod reticulate-veined .... *M. italica* (Ma, I)
2. Pod not reticulate-veined
3. Pod transversely ridged .... *M. officinalis* (Ma, I)
3. Pod concentrically striate

4. Pod blackish-brown when ripe........................... *M. infesta* (Ma)
4. Pod yellow when ripe .... *M. segetalis* (Ma)
1. Fls less than 6mm
5. Pod reticulate-veined .... *M. indica* (Ma, Mi, I)
5. Pod not reticulate-veined
6. Pod not concentrically striate
7. Corolla 5-6mm, standard and wings equal, longer than keel; pod transversely ridged, brown when ripe........... *M. officinalis* (Ma, I)
7. Corolla 4-5mm, standard and wings equal, shorter than keel; pod with transverse and S-shaped veins, brownish-yellow when ripe...... .... *M. elegans* (Ma, Mi, I)
6. Pod concentrically striate
8. Racemes of 30-50fls .... *M. segetalis* (Ma)
8. Racemes of 25 fls or less
9. Fr. globular, 3-4mm .... *M. sulcata* (Ma, Mi, I)
9. Fr. drawn out into terminal point....... *M. messanensis* (Ma, Mi, I)

MEDICAGO Medick; Melga, Mielga, Alfalfa; Schneckenklee; Minette.
1. Fls purple or blue .... *M. sativa* (Ma, Mi, I)
1. Fls yellow
2. Pnnl bush, 1-3m .... *M. arborea* (Mi, I, garden escape in Ma)
2. Annl or low pnnl
3. Coiled pod kidney-shaped, or forming less than a complete circle; fls in long pedunculate compact raceme; fr. black when ripe
4. Fls about 40 in globose head........................... *M. lupulina* (Ma)
4. Fls 3-10 in secund raceme .... *M. secundiflora* (Ma)
3. Coiled pod discoid, globular or cylindrical, with 1.5-9 complete spirals; fls rarely exceeding 8 together, usu. many less; pods brown or yellowish-brown when ripe
5. Pod diam. 12mm or more
6. Pod without spines, disc to cup-shaped
7. Pod a disc, 3-5 turns, with acute keel.. *M. orbicularis* (Ma, Mi, I)
7. Pod cup-shaped, 4-8 turns, each overlapping the one below .... *M. scutellata* (Ma, Mi, I)
6. Pod spiny, usu. globose to cylindrical .... *M. ciliaris* (Ma, I)
5. Pod diam. less than 12mm
8. Pod with strong submarginal vein (running inside the marginal vein, often apparently above and below it), visible in immature pods, sometimes becoming fused with marginal vein when mature
9. Marginal vein much weaker than submarginal vein; pod spiny .... *M. truncatula* (Ma, Mi, I)

9. Marginal and submarginal veins both well-developed at least in young pods; spines present or not
10. Grey-woolly pnnl; pods woolly, with a central hole .... *M. marina* (Ma, Mi, I)
10. Annl or pnnl, not grey-woolly, sometimes hairy and slightly grey; pod without hole
11. Pod in a lax spiral; young pods projecting from calyx as soon as petals have fallen; spines always present
12. Peduncle as long as stalk of subtending lf or longer
13. Lflets up to 25mm; pod hairless, short cylindrical .... *M. polymorpha* (Ma, Mi, I)
13. Lflets 3-6mm; pod pubescent, more or less globose............ .... *M. minima* (Ma, Mi, I)
12. Peduncles much shorter than lf stalks
14. Lflets up to 10mm; fr. a short cylinder .... *M. praecox* (Ma, Mi, I)
14. Lflets up to 25mm, usu. with a black blotch; fr. almost globular .... *M. arabica* (Ma, Mi)
11. Pod in close spiral, concealed in calyx when petals fall
15. Transverse veins not reaching marginal veins (so pod has wide veinless border)....................... *M. murex* (Ma, Mi)
15. Transverse veins reaching marginal veins (no veinless border)
16. Pod globular to elliptical................ *M. aculeata* (Ma, Mi)
16. Pod disc-shaped to cylindrical
17. Pod hairless............................ *M. littoralis* (Ma, Mi, I)
17. Pod densely glandular-hairy................ *M. rigidula* (Ma)
8. Pod without strong submarginal vein
18. Transverse veins reaching marginal vein, no veinless border; spines absent; pod cup-shaped.......... *M. scutellata* (Ma, Mi, I)
18. Transverse veins not reaching marginal vein, wide veinless border present; spines present, short and blunt .... *M. turbinata* (Ma, Mi)

TRIFOLIUM Clover, Trefoil; Trèvol; Klee; Trèfle.
1. Heads with few fls (usu. less than 6)
2. Plant hairless; petals yellow, or pink and white; all fls fertile
3. Corolla yellow................................. *T. micranthum* (Ma, Mi, I)
3. Corolla pink and white .... *T. ornithopodioides* (Mi)
2. Plant hairy; corolla of fertle fls cream-coloured, sterile fls without corolla also present .... *T. subterraneum* (Ma, Mi)

1. Most heads with more than 6 fls
 4. Calyx much inflated in fr., fruiting head becoming enlarged and soft to compress between fingers
  5. Pnnl; peduncles and lf stalks arising from creeping, rooting stems .... *T. fragiferum* (Ma, Mi, I)
  5. Annl; stems procumbent or ascending, not rooting at nodes
   6. Heads axillary; calyx hairy or woolly
    7. Fls twisted, so that standard is below wings and keel; peduncles equalling lf stalk or slightly shorter, usu. recurved in fr. .... *T. resupinatum* (Ma, Mi)
    7. Fls not twisted; peduncle much shorter than lf-stalk, not recurved in fr. .... *T. tomentosum* (Ma, Mi, I)
   6. Heads terminal; calyx not hairy or woolly .... *T. spumosum* (Mi)
 4. Calyx not, or hardly, inflated in fr.
  8. Calyx densely and conspicuously long-hairy (heads soft and woolly)
   9. Heads with broad, whitish, conspicuously veined involucre, the head falling complete with involucre in fr....... *T. cherleri* (Ma, Mi)
   9. Heads without involucre
    10. Marginal hairs of calyx and stipules dilated at base, so that the margin appears toothed .... *T. squarrosum* (Mi)
    10. Marginal hairs of calyx and stipules not dilated at base
     11. Lflets notched at tip, usu. small .... *T. stellatum* (Ma, Mi, I)
     11. Lflets long and narrow, not notched
      12. Heads numerous, terminal and axillary, up to 25mm .... *T. arvense* (Mi)
      12. Heads few, terminal, 20-80mm ...... *T. angustifolium* (Ma, Mi)
  8. Calyx hairy or not, but not conspicuously so
   13. Heads sessile or nearly so
    14. Upper surface of lvs hairless
     15. Lflets with appressed hairs beneath .... *T. bocconei* (Ma, Mi)
     15. Lflets hairless on both sides
      16. Fls white; standard shorter than calyx .... *T. suffocatum* (Ma, Mi)
      16. Fls purplish; standard 3 x calyx .... *T. glomeratum* (Ma, Mi)
    14. Lflets hairy on both sides
     17. Corolla white; lateral nerves of lflets curving backwards and thickened towards margins .... *T. scabrum* (Ma, Mi, I)
     17. Corolla pink or purplish; lateral nerves of lflets nearly straight, not thickened
      18. Fls pink, longer than calyx .... *T. striatum* (Ma)

18. Fls reddish-purple, not or scarcely longer than calyx.............. .... *T. diffusum* (Ma, Mi)

13. Heads pedunculate

19. Calyx 20 veined................................ *T. lappaceum* (Ma, Mi)

19. Calyx with up to 14 veins

20. Fls with pedicel 3mm or more, deflexed in fr.

21. Fls yellow .... *T. campestre* (Ma, Mi, I)

21. Fls creamish-white or pinkish

22. Creeping pnnl, rooting at nodes .... *T. repens* (Ma, Mi, I)

22. Annl, not rooting at nodes .... *T. nigrescens* (Ma, Mi)

20. Fls without pedicel, or with very short pedicel

23. Lvs next below head paired and opposite .... *T. squamosum* (Ma, Mi)

23. Lvs all alternate .... *T. ligusticum* (Ma, Mi)

DORYCNIUM -; Botja, Guixola, Bolaga; Backenklee; Badasse.

1. Corolla 10-20mm .... *D. hirsutum* (Ma, Mi, I)

1. Corolla 3-7mm

2. At least lower and middle stem lvs with rhachis 5mm or more........... .... *D. rectum* (Ma, Mi, I)

2. Lvs with very short rhachis or none..................... *D. pentaphyllum*

D. pentaphyllum includes here the very distinctive ssp *fulgurans* (Porta) Cardona, Llorens et Sierra (Cardona 1983), formerly known as *'Anthyllis fulgurans'* Porta. This plant does not appear in the FE under any name.

Key to ssp of *D. pentaphyllum.*

1. Spiny hedgehog plant .... ssp *fulgurans* (Ma, Mi, endemic)

1. Plant without spines

2. Heads with 5-15 flowers; calyx teeth shorter than tube............... .... ssp *pentaphyllum* (Ma, Mi, I)

2. Heads with 12-20 flowers; calyx teeth about equalling tube......... .... ssp *gracile* (Mi)

LOTUS Birdsfoot-trefoil; Corona de rei, Trébol de cuernos; Hornklee; Lotier. (The leaves are 5-foliate, but could be taken for trifoliate lvs with bracts at the base of the stalk – the English name suggests that this is how they are generally seen).

1. Lvs with 4 lflets.......................... *L. tetraphyllus* (Ma, Mi, endemic)

1. Lvs with 5 lflets

2. Pnnl

3. Plant densely silvery haired .... *L. cytisoides* (Ma, Mi, I)

3. Plant not silvery haired
  4. Lflets of upper lvs more than 3 x as long as wide .... *L. tenuis* (Ma, Mi, I)
  4. Lflets of upper lvs usu. not more than 3 x as long as wide
    5. Upper 2 calyx teeth in bud separated by an acute sinus .... *L. preslii* (?Mi)
    5. Upper 2 calyx teeth in bud separated by an obtuse sinus.. *L. corniculatus* (Ma, Mi, I)

2. Annl
  6. Pod 4-8mm diam., with dorsal groove........... *L. edulis* (Ma, Mi, I)
  6. Pod less than 2mm diam., without dorsal groove
    7. Fl. head subtended by 3 broad, egg-shaped bracts, almost as long as fls; peduncle in fr. not more than twice length of stalk of subtending lf............................... *L. ornithopodioides* (Ma, Mi, I)
    7. Bracts subtending fl. heads much shorter than fls; peduncle in fr. at least twice length stalk of subtending lf, usu. more
      8. Pod 4-6mm, not or hardly longer than calyx .... *L. parviflorus* (Ma, I)
      8. Pod 6mm or more, at least 1.5 x length calyx
        9. Pod 6-12mm, 1.5-3 x length calyx .... *L. subbiflorus* (Mi)
        9. Pod 20-30mm, 4-7 x length calyx .... *L. angustissimus* (Ma, Mi)

TETRAGONOLOBUS

Pod with wings on all 4 corners........................... *T. purpureus* (Ma, I)
Pod with wings only on 2 adjacent upper corners......... *T. requienii* (Ma)

ANTHYLLIS

1. Shrub or shrublet with woody branches
  2. Plant spiny .... *A. hermanniae* (Mi)
  2. Plant not spiny.................................. *A. cytisoides* (Ma, Mi, I)
1. Herbs, not woody
  3. Fls pink, white or lilac............................ *A. vulneraria* (Ma, Mi, I)
  3. Fls pale and deep yellow .... *A. tetraphylla* (Ma, Mi, I)

ORNITHOPUS

Heads with minute scarious bracts or none; pod slender, slightly constricted between seeds... *O. pinnatus* (Mi)
Heads subtended by a pinnate leafy bract; pod broad, flattened, not constricted between seeds................................. *O. compressus* (Ma, Mi)

CORONILLA Crown vetch; Carolina, Coronilla; Cronwicke; Coronille.

1. Small annl
   2. Upper lvs simple or trifoliate.................. *C. scorpioides* (Ma, Mi, I)
   2. Upper lvs pinnate .... *C. repanda* (Mi)
1. Robust, shrubby pnnl
   3. Stems rush-like; lvs caducous .... *C. juncea* (Ma, Mi)
   3. Stems not rush-like; lvs persisting.................. *C. valentina* (Ma, Mi)

HIPPOCREPIS Horse-shoe vetch; Desferracavalls; Hufeisenklee; Fer-à-cheval.

1. Pnnl; fls 10-15mm, numerous......... *H. balearica* (Ma, Mi, I, endemic)
1. Annl; fls 8mm or less, up to 6
   2. Peduncle about equalling lvs; fls 2-6
      3. Fls 3-5mm; pod with long papillae on seed swellings......... *H. ciliata* (Ma, Mi, I)
      3. Fls 5-8mm; pod without papillae or with very small papillae............ .... *H. multisiliquosa* (Ma, Mi)
   2. Peduncle shorter than lvs, 5mm or less; fls 1-2.............................. .... *H. unisiliquosa* (Ma, Mi, I)

SCORPIURUS

Fls 2-5, yellow .... *S. muricatus* (Ma, I)

Fls 1 (-2), orange.. *S. vermiculatus* (?BI)

HEDYSARUM

Robust, branched pnnl; fls red, in large, clover-like clusters .... *H. coronarium* (Ma, Mi)

Small annl; fls pale pinkish-purple, 2-8 together.................................. .... *H. spinosissimum* (Ma, Mi, I)

H. coronarium French honeysuckle; Anchova, Coronada, Clover, Zulla; Süssklee; Sulla.

OXALIDACEAE

OXALIS is the only genus here, and neither species is native. *O. pes-caprae* (Bermuda buttercup; Avellantes, Flor del Sueno), a common and beautiful plant, is an agricultural pest here, often found as a double form with red-tinged petals. The other is a street and garden weed.

1. Pnnl; fls bright lemon yellow; petals 20-25mm .... *O. pes-caprae* (Ma, Mi, I, introduced)
1. Annl; fls pale yellow; petals 8-10mm .... *O. corniculata* (Ma, Mi, I, introduced)

GERANIACEAE

Three genera are found here. First, cultivated in many varieties, the *Pelargonium*, commonly known as Geranium, and not so very different. Then there are the native genera, *Geranium* (Cranesbill; Rellotges; Storchschanbel; Géranium), and *Erodium* (Storksbill; Bec de cigonya; Reiherschnabel; Bec-de-grue).

The *Geranium* species here are all common in many parts of Europe, but the *Erodium* species include *E. ciconium*, a large, bluish, Southern European species, and *E. reichardii*, possibly endemic to Majorca and Minorca. It is closely related to *E. corsicum*, and possibly grows in Corsica too.

There are 5 petals and sepals, and 10 stamens (alternate stamens in *Erodium* are sterile). The fruit ends in a long beak.

KEY

Fls usu. paired; lvs palmately lobed or divided; beak of carpel rolling upwards to release seed .... *Geranium*

Fls solitary or in umbels, rarely in pairs; lvs pinnately lobed or divided, or simple and pinnately veined; beak of carpel twisting spirally when seeds ripen .... *Erodium*

GERANIUM

1. Limb of petal much longer than claw
  2. Petal deeply notched .... *G. molle* (Ma, Mi, I)
  2. Petal slightly notched, or not notched
    3. Sepals aristate (point 1-2mm)
      4. Pedicels 2-6cm .... *G. columbinum* (Ma, Mi)
      4. Pedicels 0.5-1.5cm .... *G. dissectum* (Ma, Mi, I)
    3. Sepals mucronate (point less than 0.5mm) .... *G. rotundifolium* (Ma, Mi, I)
1. Claw of petal longer than limb
  5. Lvs palmately lobed, shining .... *G. lucidum* (Ma)
  5. Lvs palmate
    6. Petals 8-12mm including claw; anthers orange or purple .... *G. robertianum* (Ma)

6. Petals 6-9mm including claw; anthers yellow .... *G. purpureum* (Ma, Mi, I)

ERODIUM

1. Fls violet; petals 10-15mm .... *E. botrys* (Ma, Mi)
1. Fls pink to lilac, occasionally bluish; petals 9mm or less
  2. Fls mostly solitary
    3. Annl; lvs pinnatifid; fls solitary or paired; petals absent or not longer than calyx .... *E. maritimum* (Is of Dragonera)
    3. Pnnl; lvs round, crenated, with cordate base; petals longer than calyx....................................... *E. reichardii* (Ma, Mi, ?endemic)
  2. Fls in umbels of 3-10
    4. Lvs pinnate at least below (sometimes pinnatisect towards apex)
      5. Lvs with small lflets alternating with larger ones; fls bluish.. *E. ciconium* (Ma)
      5. Lvs not as above; fls not bluish
        6. Lflets rarely lobed more than half way to midrib; petals more or less alike .... *E. moschatum* (Ma, Mi, I)
        6. Lvs pinnatifid or more divided; 2 upper petals often shorter and wider with dark blotches at base
          7. Pnnl; peduncle absent .... *E. acaule* (?Is Cabrera)
          7. Annl; with peduncle....................... *E. cicutarium* (Ma, Mi, I)
    4. Lvs rounded to pinnatifid or pinnatisect, with not more than 1 pair of distinct lflets
      8. Pit at apex of fr. with a wide, deep furrow beneath it .... *E. malacoides* (Ma, Mi, I)
      8. Pit at apex of fr. without furrow beneath it
        9. Base of pit gland-dotted; beak of fr. 30-40mm .... *E. chium* (Ma, Mi, I)
        9. Base of pit without glands; beak of fr. 35-90mm .... *E. laciniatum* (?Is of Cabrera)

TROPAEOLACEAE

Tropaeolum majus (Nasturtium; Murrissa, Flor de la sangre; Kapuzinerkresse; Capucine) is the only species here. It is easily recognized by its 5 orange (or in cultivated varieties dark red or yellow) petals, the spurred calyx, and large circular lvs, with the stalk attached at the centre. Native of S. America, it is widely naturalized on the banks of streams and other moist places here. (Ma, introduced)

ZYGOPHYLLACEAE

There are three species here each from a different genus. Only *Tribulus terrestris* (Caltrop, Maltese cross; Caputxina, Punxaclaus, Espigón; -; Trébol, Herse) is common. It is a rather attractive prostrate plant with small yellow flowers and neat pinnate leaves – and most unattractive spiny fruits which are adapted to be dispersed on the feet of animals, including those of the unwary bather. It often grows on the higher parts of beaches.

The other native species is *Fagonia cretica*, local names Mantell de la Vierge, Abrojo de Creta, another prickly sprawling plant with pretty 5-petalled pinkish-violet flowers.

Zygophyllum fabago is a native of SE Europe, with very distinctive leaves – there are two entire slightly fleshy leaflets, arranged on opposite sides of the stalk, without a terminal leaflet.

KEY

1. Annl; fls yellow .... *Tribulus terrestris* (Ma, Mi, I)
1. Pnnl; fls not yellow
   2. Plant erect; fls cream and orange.. .... *Zygophyllum fabago* (Ma, introduced)
   2. Plant prostrate; fls pinkish-violet................. *Fagonia cretica* (Ma, I)

LINACEAE

Herbs with simple lvs; fls with three or four petals; fruit a capsule.

Linum bienne was used in the Stone age as a source of fibre for fishing nets and traps for wild animals. It was superseded about 4,000-5,000 years ago as a crop to provide linen by *L. usitatissimum*, from Mesopotamia and the Eastern Mediterranean. Egyptian mummies were wrapped in linen bandages.

Linseed, the seed of *Linum* species, was used as a food by the Greeks and Romans, and is still eaten in parts of Europe. The oil is extracted and used in paints and varnishes.

KEY

Lvs opposite; fls white, petals and sepals 4, about 1mm .... *Radiola linoides* (Mi)

Lvs alternate; fls yellow or blue, petals and sepals 5, 4mm or more.......... .... *Linum*

LINUM Flax; Linillo, Lli; Flachs; Lin.

1. Fls blue
   2. Petals pale blue, 8-12mm .... *L. bienne* (Ma, Mi)
   2. Petals deep blue, 25-35mm .... *L. narbonense* (Ma)
1. Fls yellow
   3. Pnnl .... *L. maritimum* (Ma)
   3. Annl
      4. Lvs minutely toothed (edges rough to the touch).......................... .... *L. strictum* (Ma, Mi, I)
      4. Lvs not toothed (edges smooth) .... *L. trigynum* (Ma, Mi, I)

EUPHORBIACEAE

In this family the flowers are unisexual, and the perianth of one whorl or none. The fruit is generally a three-lobed capsule.

KEY

1. Plant with milky latex... *Euphorbia*
1. Plant without milky latex
   2. Lvs palmately lobed .... *Ricinus communis* (Ma, introduced)
   2. Lvs not palmately lobed
      3. Plant covered in star-shaped hairs........ *Chrozophora tinctoria* (Ma)
      3. Plant without any star-shaped hairs .... *Mercurialis*

Ricinus communis Castor-oil plant; Higuera del diablo, Cagamuja, Figuera infernal; Rizinus; Ricin.
Chrozophora tinctoria is called Tornasol in Catalan, Tornassol in Castilian, Tournesol in French.

MERCURIALIS Mercury; Malcoratge, Murcarol; Bingelkraut; Foivelle.
Plant densely tomentose .... *M. tomentosa* (I)
Plant sparsely hairy .... *M. annua* (Ma, Mi, I)

EUPHORBIA Spurge; Lletera, Lechetrenza; Wolfsmilch; Euphorbe.
This is a very large genus, with about thirty species in the Balearic islands. They are herbaceous plants or woody shrubs, all of which produce copious acrid and poisonous milky fluid when broken. The flowers are usually arranged in a more or less umbellate yellowish or greenish inflorescence at the top of a leafy stem.

The male flower is a single stamen on a pedicel, and the female a 3-

celled ovary on a pedicel, which enlarges and usually droops in fruit. The inflorescence is made up of several male flowers and one female grouped together in the centre of a *cyathium,* which is a ring of small greenish or yellowish glands, looking rather like petals. The glands are variously decorated. They may be oval and entire, or notched on the outer margin, or crescent moon-shaped, with the horns facing outwards, or they may have appendages of various shapes.

The decoration of the capsule and seed are important too – but if the capsule is unripe it is quite difficult to get the seed out, and if it is ripe it may burst out suddenly, with such force that you lose it altogether. It is worth trying, as it is usually difficult to be sure of the species otherwise.

Many annual spurges have a single vertical stem, with the leaves arranged alternately up it, and an umbel at the top, subtended by a ring of bracts usually rather like the stem leaves. There may be axillary branches as well as umbels, and the umbel branches may themselves be further branched, umbellately again, with a ring of bracteoles below the secondary umbel.

The perennial species tend to have several or many flowering stems, as well as non-flowering shoots, but this, the usual distinction between an annual and a perennial is rather confused by the fact that many annual spurges produce shoots at the bottom of the stem too, which flower later in the year.

1. Annl or bnnl
  2. Glands entire, not truncated or notched, nor with appendages or horns
    3. Capsule with 2 undulate wings on valves...... *E. pterococca* (Ma, Mi)
    3. Capsule unwinged
      4. Plant procumbent (maritime)..................... *E. peplis* (Ma, Mi, I)
      4. Plant erect (not maritime) .... *E. helioscopia* (Ma, Mi, I)
  2. Glands with appendages or horns, or notched or truncated
    5. Plant often prostrate; glands rounded with added appendages, not horned, where horns extend curve of gland (like crescent moon)
      6. Capsule with sharp, hairy keel on back of each valve .... *E. prostrata* (Ma)
      6. Capsule not keeled
        7. Lvs 3-7mm, entire or finely serrulate ... *E. chamaesyce* (Ma, Mi, I)
        7. Lvs 10-30mm, serrate...................... *E. nutans* (Ma, introduced)
    5. Plant usu. erect; glands horned, notched or truncated
      8. Lvs minutely serrulate .... *E. medicaginea* (Ma, Mi, I)
      8. Lvs entire
        9. Capsule 2mm or less

10. Capsule with 2 narrow wings on each valve .... *E. peplus* (Ma, Mi, I)
10. Capsule unwinged
11. Seeds hexagonal, long furrow along each face .... *E. sulcata* (Ma)
11. Seeds not as above
12. Seeds tuberculate............................. *E. exigua* (Ma, Mi, I)
12. Seeds flattened, transversely wrinkled .... *E. falcata* (Ma, I)
9. Capsule 2.5-20mm
13. Lvs opposite or almost so; capsule 8-20mm.. *E. lathyris* (Ma, introduced)
13. Lvs alternate; capsule 3.5mm or less
14. Bracteoles blunt; seeds smooth .... *E. segetalis* (Ma, Mi, I)
14. Bracteoles acute or mucronate; seeds pitted.. *E. taurinensis* (Is. Cabrera)
1. Pnnl
15. Plant very robust, commonly more than 1m
16. Shrub; infce yellow or yellowish
17. Bracteoles almost circular .... *E. dendroides* (Ma, Mi, I))
17. Bracteoles rhomboid or ovate .... *E. squamigera* (Ma)
16. Woody plant; infce green with conspicuous red-purple glands .... *E. characias* (Ma, Mi)
15. Plant usu. less than 1m
18. Glands entire, rounded; fr. tuberculate ... *E. pubescens* (Ma, Mi, I)
18. Glands truncate, notched or with horns; fr. not tuberculate
19. At least some lvs conspicuously serrate ...... *E. serrata* (Ma, Mi, I)
19. Lvs entire or minutely serrulate
20. Lvs minutely serrulate .... *E. terracina* (Ma, Mi, I)
20. Lvs entire
21. Plant with a creeping rhizome ..... *E. maresii* (Ma, Mi, endemic)
21. Plant not rhizomatous
22. Horns club-shaped
23. Stems with 2 (occasionally 3) superimposed umbels, stem more or less bare in between; lvs green .... *E. biumbellata* (Ma, I)
23. Stem with one umbel; lvs glaucous .... *E. myrsinites* (Ma)
22. Horns absent, or not club-shaped
24. Seeds pitted .... *E. segetalis* (Ma, Mi, I)
24. Seeds smooth, ridged or tuberculate
25. Lvs deflexed at base of stem.......... *E. pithyusa* (Ma, Mi, I)
25. Lvs all spreading .... *E. paralias* (Ma, Mi, I)

8
9
37
21
38
25
34
19
24
18
40
30
27
5
16
23
4
2
26
10
3
31
1
17
20
33
39
6
32
7
11
28
14
13
35
29
36
15
12
22

PLATE 2. Actinomorphic flowers with red, blue, yellow, green or white flowers with 5 petals.

	Key page
Portulacaceae	
1. *Portulaca oleracea*	42
Caryophyllaceae	
2. *Arenaria balearica*	44
3. *Cerastium glomeratum*	44
4. *Herniaria hirsuta*	45
5. *Paronychia argentea*	45
6. *Silene vulgaris*	46
Ranunculaceae	
7. *Nigella damascena*	48
8. *Ranunculus macrophyllus*	50
9. *R. aquatilis*	49
Rosaceae	
10. *Rosa arvensis*	65
11. *R. sempervirens*	65
12. *Potentilla reptans*	66
13. *Crataegus monogyna*	65
Oxalidaceae	
14. *Oxalis pes-caprae*	81
Zygophyllaceae	
15. *Tribulus terrestris*	83
Linaceae	
16. *Linum bienne*	84
17. *L. maritimum*	84
Hypericaceae	
18. *Hypericum balearicum*	93
19. *H. tomentosum*	93
Cistaceae	
20. *Cistus monspeliensis*	94
21. *C. salvifolius*	94
22. *Tuberaria guttata*	95
23. *Fumana thymifolia*	95
Cucurbitaceae	
24. *Ecballium elaterium*	96
Umbelliferae	
25. *Naufraga balearica*	100
26. *Foeniculum vulgare*	100
27. *Bupleurum barceloi*	104
Primulaceae	
28. *Cyclamen balearicum*	106
29. *Anagallis arvensis*	107
Gentianaceae	
30. *Centaurium maritimum*	110
Asclepiadaceae	
31. *Vincetoxicum hirundinaria*	111
Convolvulaceae	
32. *Calystegia sepium*	115
Boraginaceae	
33. *Heliotropium europaeum*	117
34. *Myosotis arvensis*	116
35. *Cynoglossum creticum*	119
36. *C. cheirifolium*	119
37. *Anchusa azurea*	117
38. *A. arvensis*	117
39. *Borago officinalis*	115
Scrophulariaceae	
40. *Sibthorbia africana*	128

RUTACEAE

RUTA (Rue; Ruda; Raute; Rue) is the only native genus. Citrus fruits, cultivated here, are Rutaceae too. *Ruta* species are slightly shrubby plants, with much-divided often glaucous and strong-smelling leaves. The plants were used medicinally, and in classical times Ruta was a name for any bitter herb, and hence any unpleasantness.

The yellow flowers usually have four petals and sepals and 8 stamens, though in some species flowers occurring at divarications of the inflorescence have 5 petals and sepals and 10 stamens.

1. Petals fringed with long hairs
   2. Stem glandular hairy above .... *R. angustifolia* (Ma, Mi, I)
   2. Stem hairless .... *R. chalepensis* (Ma, Mi, I)
1. Petals not fringed .... *R. montana* (Ma)

CNEORACEAE

Cneorum tricoccon (Escanyacabres, Olivella in Majorca, Raspall in Ibiza, Camélée tricoque in France) is a small shrub with narrow leathery leaves and yellow three-petalled flowers. The fruit is 3-lobed, turning from green through yellow to red and finally black. (Ma, Mi, I)

POLYGALACEAE

POLYGALA (Milkwort; Hierba lechera; Kreuzblume; Polygala) is the only genus here, all herbs or very small shrubs. The flowers have three small petals united to a tube formed from the fused stamens, which are usually brightly coloured, all enclosed in a pair of almost transparent enlarged inner sepals. The ovary is superior, with a simple style.

POLYGALA

1. Little branched annl; fls greenish-white........ *P. monspeliaca* (Ma, Mi)
1. Much branched pnnl
   2. Lf margins inrolled; stamens bright pinkish-purple; inner sepals greenish-white .... *P. rupestris* (Ma, Mi, I)
   2. Lf margins not inrolled; fls blue, pink or white .... *P. vulgaris* (Ma)

CORIARIACEAE

Coriaria myrtifolia (Emborachacabras in Castilian) belongs to the only genus in the family, which has an interesting discontinuous distribution. Species are only found in New Zealand, Borneo and the Himalayas, parts of SW and Central America, and the Mediterraneaen. The species here is a shrub, up to 3m high, with leathery oval leaves and clusters of 5-lobed purple fruit, turning black. (I)

ANACARDIACEAE

This family includes the trees which produce Cashew and Pistacia nuts. Here there is one genus, *Pistacia*, and two species. *P. terebinthus* (Turpentine tree; Cornicabra, Cornizuelo, Mataselva,Púdol; Terebinthe; Térebinthe) is a rare shrub of less accessible parts of Majorca; *P. lentiscus* (Mastic tree; Lentisco, Matot, Mata roja; Mastixbaum: Arbre-au-mastic) is common in dry places such as maritime dunes. It is commonly a shrub up to about 3m high, but reaches 8m on the coast of SE Majorca. They both produce sweet-smelling gums used in medicine and dentistry, and in chewing-gum.

PISTACIA

Lf pinnate, no terminal lflet, axis winged............ *P. lentiscus* (Ma, Mi, I)
Lf pinnate with terminal lflet, axis unwinged .... *P. terebinthus* (Ma)

ACERACEAE

Only *Acer granatense* here, a Maple (Acirón; Ahorn; Erable) known only from Majorca and a small area of mainland Spain. The leaves are five-lobed, the three upper lobes with almost parallel sides. It grows in the mountains of Majorca, usually stunted by wind (and goats). (Ma)

AQUIFOLIACEAE

Ilex aquifolium (Holly; Arbre de visc, Grévol, Gorostiya, Agrifolio; Stechpalme: Houx) occurs here in mountain crevices. It is not common. (Ma)

BUXACEAE

The only species here is *Buxus balearica*, confined to Majorca, a small area of mainland Spain, and N. Africa. There is a very fine mature specimen in Cambridge University Botanic Garden. It is like *Buxus sempervirens* (Box; Boj; Buchsbaum; Buis), but the styles are longer and the leaves larger and a brighter green. It was formerly common in Majorca, but woodcutters and charcoal burners destroyed most of it. (Ma)

RHAMNACEAE

This family is better known in Northern Europe by *Rhamnus catharticus* (Buckthorn; Espino cerval; Kreuzdorn; Nerprun), which is not found here, though common in most of Europe, and by the cultivated species of *Ceanothus*, with panicles of (usually) powdery blue flowers. There are three species of *Rhamnus* here, one of them endemic. The leaves are simple and leathery, and the flowers small and inconspicuous in axillary clusters, with four or five yellow petals.

RHAMNUS

1. Spines present .... *R. lycioides* (Ma, I)
1. Spines absent
   2. Lf margins prickly.......... *R. ludovici-salvatoris* (Ma, Mi, I, endemic)
   2. Lf margins entire, or with a few teeth, not prickly ....
   .... *R. alaternus* (Ma, Mi, I)

Local names: *R. alaternus* Llampúgol, Mataselva, Coscollina, Aladern.
R. lycioides Alicantius, Espinalera, Arot, Espino negro.
R. ludovici-salvatoris Rotaboc, Llampúdol bord.

VITACEAE

Vitis vinifera (Grapevine; Vid, Parra; Weinrebe: Vigne) has been cultivated here from Roman times. It is often found growing wild either in the cultivated form or in a wild form with smaller fruit. *Phylloxera* resistant American species have also been introduced in recent times as stock for grafting, and these may be found growing apparently wild too.

MALVACEAE

The name Malva for plants of this family was used by Pliny and Cicero long before Linnaeus (1707-1778, the celebrated Swedish botanist who invented the binomial system for naming plants). In Roman times the name was probably mainly used for *Malva sylvestris* and *Lavatera arborea*.

The French and thence the English word mauve is derived from Malva, and mauve is the characteristic colour of many species, and of nearly all species here, though the exact shade varies. Some members of the family have a capsule as fruit, but in all except *Kosteletskya pentacarpos* here the fruit is like a round cheese divided into separate portions, each portion being a separate nutlet. The fruit is called Katzenkäse (cat cheese) in German. Many species are cultivated here, particulary *Hibiscus* species.

KEY

1. Carpels 3-5; fr. a capsule .... *Kosteletskya pentacarpos* (Mi)
1. Carpels 6 or more; fr. a whorl of nutlets
  2. Epicalyx of 2-3 segments
    3. Epicalyx segments free to the base .... *Malva*
    3. Epicalyx segments united below .... *Lavatera*
  2. Epicalyx of 6-9 segments .... *Althaea hirsuta* (Ma, Mi, I)

MALVA Mallow; Malva; Malve; Mauve.

1. Petal with bearded claw, veins darker than background .... *M. sylvestris* (Ma, Mi, I)
1. Petal claw almost or completely hairless; veins not or hardly darker than background
  2. Fr. pedicel up to 1cm; petals 4-5mm......... *M. parviflora* (Ma, Mi, I)
  2. Fr. pedicel usu. 1cm or more; petals 10-12mm.............................. .... *M. nicaeensis* (Ma, Mi)

LAVATERA

1. Fls in clusters
  2. Annl or bnnl, entirely herbaceous .... *L. cretica* (Ma, Mi, I)
  2. Pnnl, woody at base
    3. Plant with glandular hairs, smelling of musk.............................. .... *L. triloba* ssp *pallescens* (Mi, endemic ssp)
    3. Plant without glandular hairs or characteristic smell...................... .... *L. arborea* (Ma, Mi, I, introduced)
1. Fls solitary
  4. Annl; petals more or less overlapping at base

5. Fls 5-7cm .... *L. trimestris* (Ma, Mi, I)
5. Fls 1.5-3cm .... *L. punctata* (Ma, Mi)
4. Pnnl; petals widely separated at base, sepals visible between
6. Fls sessile or nearly so .... *L. olbia* (Ma, Mi)
6. Fls long pedicellate.............................. *L. maritima* (Ma, Mi, I)

THYMELAEACEAE

Thymelaea and *Daphne* are the genera here. Both have a tubular flower with four spreading petal-like lobes. The plants are mostly shrubby with simple leaves.

KEY

Fr. dry, enclosed in persistent calyx; fls yellow or greenish- yellow .... *Thymelaea*
Fr. fleshy; calyx not persistent; fls purplish-yellow or white .... *Daphne*

THYMELAEA

1. Annl... *T. pubescens* (Ma, Mi, I)
1. Pnnl, dwarf shrubs
2. Leaves glabrous.. *T. dioica* (I)
2. Lvs hairy
3. Lvs bright green and hairless beneath, white tomentose above........ .... *T. hirsuta* (Ma, Mi, I)
3. Lvs grey and hairy on both sides.... *T. myrtifolia* (Ma, Mi, endemic)

DAPHNE The name Laurus was applied in classical times to almost any bush with dark green leaves, including *Laurus nobilis, Myrtus communis, Viburnum tinus* and *Daphne* species. When the nymph Daphne, chased by Apollo, changed herself into a Laurel tree, she may have chosen any of them; but later tradition gave *Laurus nobilis* the name, and *Daphne* the nymph.

Fls terminal, purplish-yellow; lvs fringed with hairs............................ .... *D. rodriguezii* (Mi, endemic)
Fls in terminal branched clusters; fls white .... *D. gnidium* (Ma, Mi, I)

D. gnidium is known locally as Matapoll, Tintorell or Torvisco, in French as Saint-bois or Garou.

GUTTIFERAE

The only genus here is *Hypericum* (St John's Wort; Périco, Hierba de Sant Juan; Johanniskraut; Millepertuis), a genus easy to recognize because the species all have very similar flowers. The ovary is superior, and there are many stamens, sometimes united into 3 or 5 bundles. The leaves are opposite in the species here. Parts of the plant are sometimes dotted with black or whitish glands.

HYPERICUM

1. Erect shrubs to 1m or more
  2. Lvs with yellowish-white glandular vesicles on margins .... *H. balearicum* (Ma, Mi, I, endemic)
  2. Lvs without marginal vesicles .... *H. hircinum* (Ma)
1. Herbs or low, often procumbent, shrubby plants
  3. Leaves tomentose .... *H. tomentosum* (Ma, I)
  3. Lvs hairless
    4. Sepals without black glands .... *H. perforatum* (Ma, Mi, I)
    4. Sepals with black glands
      5. Plant usu. much-branched and rooting at the base; lvs minutely serrated .... *H. australe* (Mi)
      5. Stems erect, usu. branching only in terminal infce; lvs entire .... *H. perfoliatum* (Ma, Mi)

VIOLACEAE

Viola (Violet; Violetta; Veilchen; Violette) is the only genus here. Botanists differ about which species occur here, and the key here is derived partly from Bonafè's account and partly from Duvigneaud's. I have used Bonafè's descriptions as well as those in the FE.

1. Plant woody, at least below .... *V. arborescens* (Ma, Mi, I)
1. Plant herbaceous
  2. Stipules broad, triangular or ovate
    3. Leaves hairless, very shiny .... *V. jaubertiana* (Ma, endemic)
    3. Lvs hairy or hairless, not conspicuously shining .... '*V. stolonifera*' (Ma, Mi, I)
  2. Stipules linear to lanceolate
    4. Stolons short, stout....................................... *V. suavis* (Ma, I)
    4. Stolons long and slender, or absent ..... *V. alba* ssp *dehnhardtii* (Ma)

V. suavis includes 'ssp *segobricensis*' (Pau) Becker and 'ssp *sepincola* (Jord.) Becker var *barceloi*' O. de Bolòs et J. Vigo. *'Viola stolonifera'* Rodr. is equated in FE with *V. odorata*. Duvigneaud places it in the *V. odorata* group.

CISTACEAE

Many of the popular names for this family include the word 'Rose'.
The flowers are like roses, regular and five-petalled, with the same colours, pink, white and yellow, and with numerous yellow stamens.

But there is one obvious difference – there is only one style. Added to this, the petals are often crumpled, like the petals of a Poppy, and the plants are not prickly or spiny. The leaves are simple and usually opposite, not pinnate and alternate like rose leaves, and the fruit is a dry capsule, not a fleshy rose-hip.

But they can still look very like roses.

KEY

1. Carpels 5; all sepals more or less the same length .... *Cistus*
1. Carpels 3; inner 3 sepals much longer than outer 2
  2. Style 0, or very short
    3. Shrub, to 60cm or more .... *Halimium halimifolium* (Ma)
    3. Annl or pnnl herb to 30cm, usu. less .... *Tuberaria*
  2. Style elongated, and bent or curved
    4. Stamens all fertile; lvs opposite .... *Helianthemum*
    4. Outer stamens sterile; upper lvs usu. alternate................ *Fumana*

CISTUS Sunrose; Jaguarzo, Estepa; Sonnenröschen, Sandröschen; Ciste.

1. Sepals 5
  2. Style as long as stamens; petals pink to pinkish-mauve
    3. Lvs sessile, with 3 roughly parallel veins .... *C. albidus* (Ma, Mi, I)
    3. Lvs petiolate, pinnately veined............... *C. incanus* (Mi, ?Ma, ?I)
  2. Stigma sessile or almost so; petals white
    4. Lvs almost sessile, almost linear, shining ....
    .... *C. monspeliensis* (Ma, Mi, I)
    4. Lvs stalked, oval, sage-like (not shining) ... *C. salvifolius* (Ma, Mi, I)
1. Sepals 3 .... *C. clusii* (Ma, I)

TUBERARIA

Pnnl; fls 20-30mm, not blotched .... *T. lignosa* (Mi)

Annl; fls up to 12mm, petal often dark-blotched .... *T. guttata* (Ma, Mi, I)

HELIANTHEMUM Rockrose; Hierba del quadrillo, Perdiguera; Sonnenröschen; Hélianthème.

1. Fls pink.. *H. appeninum* (Ma)
1. Fls yellow
 2. Lvs densely grey hairy or white felted
  3. Annl; calyx moderately hairy........................ *H. salicifolium* (Ma)
  3. Pnnl; calyx densely covered in long white hairs ... *H. caput-felis* (Ma)
 2. Lvs green, only slightly hairy.................... *H. origanifolium* (Ma, I)

FUMANA

1. Lvs roughly equally spaced on stems, not or hardly smaller above
 2. Plant procumbent; fruiting pedicels recurved from base,not longer than adjacent lvs .... *F. procumbens* (Ma)
 2. Plant erect or straggling; fruiting pedicels much longer than adjacent lvs, spreading with deflexed apex .... *F. ericoides* (Ma, Mi, I)
1. Lvs unequally spaced on stem, more or less abruptly reduced above to form small bracts in the infce
 3. Lvs opposite.. *F. thymifolia* (Ma, Mi, I)
 3. Lvs alternate .... *F. laevipes* (Ma, Mi, I)

TAMARICACEAE

TAMARIX (Tamarisk; Atarce; Tamarisk; Tamarice) is the only genus here. They are small graceful trees with tiny scale-like leaves and slender, drooping racemes or spikes of very small pink or whitish flowers.

They are often planted for shade around the beaches, where they seem to be very salt-tolerant. It is often impossible to tell whether a particular tree is natural or planted.

The petals fall easily, and seem difficult to count; other features should probably be preferred in this key.

TAMARIX

1. Sepals and petals 4; racemes 50-150 x 8-9mm; bracts much longer than fls.. *T. boveana* (Ma)
1. Sepals and petals 5; racemes not more than 60 x 8mm; bracts not longer than fls.
 2. Petals 2-3mm, at least some persistent; racemes 5-8mm wide........... .... *T. africana* (Ma, Mi, I)

2. Petals 1.5-2mm, all caducous; racemes up to 5mm wide
   3. Bracts entire; sepals finely toothed .... *T. canariensis* (Ma, I)
   3. Bracts more or less toothed; sepals entire .... *T. gallica* (Ma, Mi)

FRANKENIACEAE

FRANKENIA is the only genus here, prostrate plants of saline soils, with small (6mm or less) actinomorphic, pink, white or violet flowers.

1. Annl; fls pale or deep violet, crowded in short terminal and axillary spikes; lvs usu. flat .... *F. pulvurulenta* (Ma, Mi, I)
1. Pnnl; fls purplish-pink to white, not in spikes; lvs usu. inrolled
   2. Fls solitary, sessile in forks of the branches or terminating short lateral shoots .... *F. laevis* (Ma, Mi, I)
   2. Fls in dense, terminal branched clusters.......... *F. hirsuta* (Ma, Mi, I)

ELATINACEAE

Elatine is the only genus in Europe. *E. macropoda*, the plant here, is completely submerged in fresh water. The leaves are opposite, and the flowers axillary, with four petals and sepals, the petals pale red and shorter than the sepals. The fruit is a globose capsule containing straight and elongated, or asymmetrically horseshoe-shaped seeds. (Ma, Mi)

CUCURBITACEAE

Ecballium elaterium (Squirting cucumber; Esquitxagassos, Pepino del diablo; Spritzgurke; Momordique, Giclet) is the only species growing wild here. The plants are prostrate, fleshy and rough-haired. The fruits are egg-shaped, about the size of a small hen's egg, pendant from the top of a vertical, rigid pedicel arising from a leaf axil. The seeds inside are under great pressure – be careful if you open the fruit! The same family gives us cucumbers (Pepino; Gurke; Concombre), Melons (Mélo; Melone; Melon) and Pumpkins (Calabaza; Kürbis; Courge) and many other edible and ornamental plants. (Ma, Mi, I)

CACTACEAE

Opuntia ficus-indica (Prickly pear; Figuerassa, Figuera de moro, Chumbera, Nopaleera; Feigenkaktus; Figuier de Barbarie), a native of Tropical

America, is well established here, and others are cultivated and occasionally found more or less wild. It is a massive sprawling plant, woody below, with large dinner-plate sized discs growing one out of another, thick, fleshy, glaucous and very prickly.

The discs are thickenings in the stem, and the leaves are reduced to prickles. Yellow flowers are produced round the margins of the upper discs, and these are followed by sweet fruits – edible, but not sufficiently so to repay the discomfort of peeling one. (Ma, Mi, I, introduced)

LYTHRACEAE

LYTHRUM (Loosestrife; Hierba del toro, Blavet; Weiderich). The species here are straggling plants of damp places with simple leaves, opposite, alternate or spirally arranged along the soft stems. Each of the upper leaves has a more or less sessile flower in the axil. The flowers have 4-6 carmine red petals, or sometimes no petals at all. The differences between the species are often small.

LYTHRUM

1. Pnnl; petals 5-6mm; stamens 12, some longer than petals ... *L. junceum* (Ma, Mi, I)
1. Annl; petals up to 4mm, usu. less; stamens rarely more than 6, not longer than petals
   2. Petals minute and caducous or absent; calyx tube broadly bell-shaped .... *L. borysthenicum* (Ma)
   2. Petals at least 1mm, usu. more; calyx tube narrow funnel-shaped, tapering to base
      3. Stems 10-25cm; most lvs more than 2mm wide; petals 5-6; stamens usu. 6 .... *L. hyssopifolia* (Ma, Mi, I)
      3. Stems 5-12cm; lvs less than 2mm wide; petals usu. 4; stamens usu. 2 .... *L. thymifolia* (Mi)

TRAPACEAE

Trapa natans is a floating herb of SE Europe and the warmer parts of Asia. It has two kinds of leaf, one submerged and root-like, and one floating, more or less rectangular. The floating leaves have long stalks each with a spindle-shaped swelling, and are arranged in rosettes, from which arise the white, four-petalled flowers, followed by edible spiny fruit. (?Ma, introduced)

MYRTACEAE

This family includes the *Eucalyptus* (Gum) trees of Australia and Tasmania, which are planted here as shade plants. Many species thrive and multiply.

The trees are graceful and usually slender, often with peeling or shredding bark and leathery glaucous leaves, which make a pleasant rattling noise when the wind blows. The flowers are creamish-white with petals and sepals which harden in the bud to form a cap-like operculum which falls off, leaving a mass of stamens surrounding a circular, greenish pool of sticky nectar much sought after by bees.

There is a native genus too, with *Myrtus communis* (Myrtle; Mirto; Myrte; Myrte) the only species here. This is a small shrub with pleasantly aromatic dark, shiny leaves, and fragrant white axillary flowers very like those of a gum tree except for the persistent sepals and petals. (Ma, Mi, I)

PUNICACEAE

Punica granatum (Pomegranate; Magraner, Granado; Granatbaum; Granadier) is often cultivated in the area, usually as a single tree rather than a field crop. It is more attractive for its brilliant scarlet, 5-7-petalled large flowers than for the rather tasteless pippy fruit, so much painted by Renaissance painters. When fully grown this is a small tree, but wayside shrublets flowering when only 50cm or so high are common, probably from discarded pips. (Ma, introduced)

ONAGRACEAE

The only native genus here is *Epilobium*; *Oenothera rosea* is recorded as an introduced plant in Majorca, but does not seem to be common. It is very similar to *Epilobium*, but without the long plumed seeds. *Epilobium* does have plumed seeds, and a terminal cluster of 4-petalled rose-pink flowers with a swollen white stigma.

EPILOBIUM

1. Fls 15-23mm .... *E. hirsutum* (Ma, Mi)
1. Fls 6-9mm
   2. Stigma with 4 spreading lobes .... *E. parviflorum* (Ma, Mi, I)
   2. Stigma entire .... *E. tetragonum* (Ma, Mi)

HALORAGACEAE

Myriophyllum is the only genus here – two species of submerged aquatic plants with flowering spikes of rather inconspicuous reddish or greenish flowers protruding from the water. The upper flowers in the spike are usually male, with small petals that soon fall off, and the lower female, with even fewer petals or none at all. The submerged leaves are in whorls, and are pinnate with thread-like segments.

Lvs usu. 5 in a whorl, much longer than the distance between whorls; lf-like bracts at least as long as fls right to tip of spike *M. verticillatum* (Ma)

Lvs usu. 4 in a whorl, about equalling distance between whorls; bracts towards tip of spike entire and shorter than fls......... *M. spicatum* (Ma, Mi)

THELIGONACEAE

Theligonum cynocrambe, the only species here, is a rather negligeable plant, for enthusiasts only. It isn't even rare. It is annual, slightly succulent and foetid, with insignificant green flowers, male and female separate in axillary clusters. The leaves are entire and stalked, the lower opposite, the upper alternate. Superficially it looks rather like a fleshy form of *Parietaria officinalis*. (Ma, Mi, I)

ARALIACEAE

The one species here, needing no description, is *Hedera helix* (Ivy; Heura, Hiedra; Efeu; Lierre). It must be as familiar to city as to country people, covering buildings and trees with its dark, shiny evergreen leaves with complete impartiality. (Ma, Mi, I)

UMBELLIFERAE

This family is usually easily recognized by the presence of an *umbel*, like an umbrella, with spokes. The spokes are a framework of stalks or rays, all coming from the same point, each usually with a mini-umbel with raylets at the end.

The umbel may sometimes be more globular than an umbrella, or sometimes flatter.

In many species there are bracts at the point where the rays converge and bracteoles where the raylets converge.

Individual flowers at the tips of the rays or raylets are five-petalled, often with the tip of each petal incurved. The fruit is two dry carpels suspended on either side of an axis, united at first but usually separating into two when ripe.

A few species are not easily recognized as members of the family. The least obvious members are species of the rather thistle-like *Eryngium*.

KEY

1. Plant very small, up to 4cm; basal lvs 5-10mm, 3-foliate or pinnate, entire, or with the terminal lflet lobed.. *Naufraga balearica* (Ma, otherwise only known from Corsica)
1. Plant normally much larger (occasionally seedlings produced late in the season flower when very small, but lvs not as above).
  2. Plant with conspicuously spiny leaves
    3. Fls greenish or bluish in dense, not obviously umbellate, heads .... *Eryngium*
    3. Fls white, in lax umbel, rays 5-8.... *Echinophora spinosa* (Ma, Mi, I)
  2. Lvs not conspicuously spiny, teeth sometimes slightly prickly
    4. Fls yellow or greenish-yellow
      5. All lvs entire.. *Bupleurum*
      5. At least some lvs toothed, lobed or more deeply divided
        6. Ultimate lf segments thread-like
          7. Huge plant (1-5m); terminal umbel of female fls overtopped by smaller, lateral umbels of male fls... *Ferula communis* (Ma, Mi, I)
          7. Plant not usu. exceeding 130cm; fls male or hermaphrodite
            8. Bracteoles present; fr. with 4 wide and 2 narrow wings .... *Elaeoselinum asclepium* (I)
            8. Bracteoles usu. absent; fr. unwinged or with single wing
              9. Fr. 4mm or more
                10. Fr. unwinged, scarcely compressed................................ .... *Foeniculum vulgare* (Ma, Mi, I)
                10. Fr. with a single wing, strongly compressed dorsally........... .... *Anethum graveolens* (?Ma)
              9. Fr. not more than 2.5mm .... *Ridolfia segetum* (Ma, I)
        6. Ultimate lf segments not thread-like, sometimes narrow linear
          11. Bracts and bracteoles numerous
            12. Fleshy plant; fls yellowish-green (maritime) .... *Crithmum maritimum* (Ma, Mi, I)
            12. Plant not fleshy; fls bright yellow ..... *Kundmannia sicula* (Ma)
          11. Bracts rarely more than 3; bracteoles up to 5

13. Lvs broad, coarse, shining, upper and basal entire, lower cauline pinnate, with entire or shallowly lobed segments .... *Pastinaca lucida* (Ma, Mi, endemic)
13. Lvs much more divided, basal lvs never entire
14. Lateral wings of fr. 2-6mm broad .... *Thapsia*
14. Fr. unwinged
15. Umbels almost spherical; fr. 8mm .... *Smyrnium olusatrum* (Ma, Mi, I)
15. Umbels flat-topped; fr. 2.5mm .... *Petroselinum crispum* (Ma, Mi, I)
4. Fls white or pinkish-white
16. Ripe fr. spiny
17. Outer carpel spiny, inner tuberculate .... *Torilis*
17. Both carpels equally spiny
18. Fr. less than 4mm
19. Fr. with well-developed beak, up to 2mm............................ .... *Anthriscus caucalis* (Ma)
19. Fr. not beaked .... *Torilis*
18. Fr. 6mm or more
20. Bracts pinnatisect or bi-pinnatisect .... *Daucus carota* (Ma, Mi, I)
20. Bracts linear or trifid or absent .... *Pseudorlaya pumila* (Ma, Mi, I)
16. Ripe fr. not spiny
21. Fr. hairy or tomentose or pubescent with star-shaped hairs
22. Umbels simple (raylets absent) .... *Bowlesia incana* (Ma)
22. Umbels compound (raylets present)
23. Fr. 2mm .... *Pimpinella*
23. Fr. 5mm or more
24. One petal of outer fls 2-lobed and larger than others, like 2 large petals; seed unwinged...... *Tordylium apulum* (Ma, Mi)
24. Marginal fls not differing from others; seeds unwinged .... *Magydaris panacifolia* (Ma, Mi)
21. Fr. neither hairy nor tomentose
25. Fr. 15-18mm long.............. *Scandix pecten-veneris* (Ma, Mi, I)
25. Fr. 7mm or less
26. Fr. winged, sometimes narrowly
27. Rays 11-16; bracts usu. absent; seeds 4-4.5mm .... *Ligusticum lucidum* (Ma, endemic ssp *huteri* here)
27. Rays 20-50; bracts usu. numerous; seeds 6mm .... *Laserpitum gallicum* (Ma)

PLATE 3. Actinomorphic flowers with 3, 4 or 6-10 petals.

	Key page
Santalaceae	
1. *Osyris alba*	34
Ranunculaceae	
2. *Adonis annua*	49
3. *Clematis flammula*	49
4. *C. cirrhosa*	49
5. *Ranunculus ficaria*	50
Papaveraceae	
6. *Papaver pinnatifidum*	52
7. *P. hybridum*	52
8. *Glaucium flavum*	52
Cruciferae	
9. *Sisymbrium orientale*	59
10. *Matthiola sinuata*	59
11. *Lobularia maritima*	56
12. *Moricandia arvensis*	58
Rutaceae	
13. *Ruta angustifolia*	78
Cneoraceae	
14. *Cneorum tricoccon*	
Thymelaeaceae	
15. *Daphne gnidium*	92
16. *Thymelaea hirsuta*	92
17. *T. myrtifolia*	92
Lythraceae	
18. *Lythrum junceum*	97
Onagraceae	
19. *Epilobium hirsutum*	98
Ericaceae	
20. *Erica arborea*	105
21. *E. multiflora*	105

	Key page
Gentianaceae	
22. *Blackstonia perfoliata*	110
23. *Centaurium pulchellum**	110
Rubiaceae	
24. *Sherardia arvensis*	112
25. *Asperula cynanchica*	113
Plantaginaceae	
26. *Plantago lagopus*	133
27. *P. albicans*	135
Liliaceae	
28. *Asphodelus aestivus*	160
29. *Ornithogalum narbonense*	160
30. *O. arabicum*	160
31. *Muscari comosum*	161
32. *M. neglectum*	161
33. *Asparagus acutifolius*	162
34. *Smilax aspera*	160
35. *Allium roseum*	161
36. *A. nigrum*	161
Amaryllidaceae	
37. *Leucojum aestivum* ssp *pulchellum*	163
38. *Narcissus serotinus*	163
39. *N. tazetta*	163
40. *Pancratium marinum*	163
Iridaceae	
41. *Iris pseudacorus*	164
42. *Romulea columnae* ssp *columnae*	165

* *Centaurium pulchellum* is also illustrated as a flower with 5 petals. Usually only small plants with a single flower (a common form) have 4-petalled flowers.

23
14
16
13
40
36
17
11
15
12
9
18
28
1
2
19
10
30
29
35
5
31
41
32
6
4
34
33
25
26
7
39
3
37
38
21
24
27
42
22
8
20

26. Fr. unwinged
 28. Bracts lf-like or variously divided
  29. Umbels lf-opposed (plant growing in water) .... *Berula erecta* (Ma, Mi, I)
  29. Umbels terminal (plant not growing in water) .... *Ammi*
 28. Bracts neither lf-like nor divided, sometimes absent
  30. Stems usu. purple-spotted; bracteoles 3-6, on outside of umbel only .... *Conium maculatum* (Ma, Mi, I)
  30. Stem not purple-spotted; bracteoles absent, or evenly distributed
   31. Styles 0.5 x length of fr. or more .... *Oenanthe*
   31. Styles less than 0.5 x length fr., or absent
    32. Bracts 5-10 .... *Bunium*
    32. Bracts 0-4
     33. Fr. 4.5-7mm........................ *Bifora testiculata* (Ma, I)
     33. Fr. 3mm or less.. *Apium*

Echinophora spinosa Zanahorıa marina (Castilian), Panais porc-épines in French.
Ferula communis Giant fennel; Canyaferla, Cañaheja;-; Férule.
Foeniculum vulgare Fennel; Fonoll, Hinojo; Fenchel; Fenouil.
Anethum graveolens Dill; Fonoll pudent, Eneldo; Dill; Aneth.
Crithmum maritimum Rock samphire; Fonoll mari, Crëte marina, Peregil del mar;-; Criste-marine.
Pastinaca lucida: local names Figuera infernal, Carnosa, Carnassa.
Petroselinum crispum Parsley; Perexil, Perejil; Petersilie; Persil.
Smyrnium olusatrum Alexanders; Cugul, Apiocaballar, Cañarejo; Gelbolde; Maceron.
Anthriscus Cow-parsley; Cerfeuil de fous, Ahogo gatos, Perifollo; Kerbel; Persil sauvage.
Daucus carota Wild carrot; Zanahoria, Botges, Cenoria; Möhre; Carotte sauvage.
Scandix pecten-veneris Shepherd's needle; Agullot, Agulles de pastor; Nadelkerbel; Peigne de Vénus.
Conium maculatum Spotted hemlock; Perejil lobuno, Ceguda mayor; Schierling; Grande ciguë.

ERYNGIUM
 Plant glaucous; bracts with stout spiny teeth..... *E. maritimum* (Ma, Mi, I)
 Plant green; bracts nearly or quite entire .... *E. campestre* (Ma, Mi, I)

BUNIUM

Sepals absent or minute; fruiting pedicels much narrower than fr. .... *B. bulbocastanum* (Ma, Mi, I)

Sepals distinct; fruiting pedicels almost as broad as fr. .... *B. pachypodum* (Ma)

PIMPINELLA

Bracteoles absent or few; fr. 2mm, shortly tomentose .... *P. tragium* (Ma)

Bracteoles usu. 5; fr. 5-7mm, not tomentose .. *P. bicknellii* (Ma, endemic)

OENANTHE

Stem hollow; fr. 5mm, globular .... *O. globosa* (Ma, Mi)

Stem solid (or with small central cavity when old); fr. 2-3mm, ovoid........ .... *O. lachenalii* (Ma, Mi)

BUPLEURUM

1. Upper lvs perfoliate .... *B. lancifolium* (Ma, Mi, I)
1. Upper lvs not perfoliate
   2. Pnnl
      3. Leaves stalked (stalk amplexicaul at base); rays 2-5 ... *B. rigidum* (I)
      3. Leaves not or hardly stalked; rays 5-8 (-12) .... *B. barceloi* (Ma, I, endemic)
   2. Annl
      4. Margins of bracts and bracteoles narrowly scarious........................ .... *B. baldense* (Ma, Mi)
      4. Margins of bracts and bracteoles not scarious .... *B. semicompositum* (Ma, Mi, I)

APIUM

1. Plant minute .... '*A. bermejoi*' Ll Llorens (Mi, endemic)
1. Plant not minute
   2. Bracteoles absent .... *A. graveolens* (Ma)
   2. Bracteoles 5-7...................................... *A. nodiflorus* (Ma, Mi, I)

A. graveolens Celery; Apio, Apit; Sellerie; Céleri.

AMMI

Rays up to 150, becoming erect, thickened and hardened in fr............... .... *A. visnaga* (Ma)

Rays 15-60, slender in fl. and fr. .... *A. majus* (Ma, Mi)

THAPSIA

Plant pubescent; terminal segments of leaves toothed .... *T. villosa* (Ma, Mi)

Plant hairless or slightly bristly; terminal segments of lvs entire or with up to 2 teeth .... *T. garganica* (I)

TORILIS Hedge parsley;-; Klettenkerbel; Torilis.

1. Umbels mostly lf-opposed, peduncles up to 5cm, usu. less
   2. Rays very short, concealed by fls or fr. .... *T. nodosa* (Ma, Mi, I)
   2. Rays easily visible .... *T. leptophylla* (Ma)
1. Umbels mostly terminal, peduncles exceeding 5cm .... *T. arvensis* (Ma, Mi, I)

PYROLACEAE

Monotropa hypopitys is the only species here, a pinkish-white or ochre saprophyte without chlorophyll. At first sight it could be taken for *Orobanche* or a saprophytic orchid, but unlike these the flowers are actinomorphic. There may be four or five petals. (Ma)

ERICACEAE

Two genera occur here, *Erica* and *Arbutus*, both with a tubular corolla with short lobes.

KEY

Lvs 4-10cm, elliptical, shiny dark green above and pale green below; corolla lobes 5; fr. a fleshy berry .... *Arbutus unedo* (Ma, Mi, I)

Lvs up to 11mm; corolla lobes 4; fr. a dry capsule .... *Erica*

ERICA Heather; Bruc, Brezo; Heide; Bruyère, Brande.

1. Fls 2-3mm, white or greenish
   2. Fls white, in dense pyramidal clusters.............. *E. arborea* (Ma, Mi)
   2. Fls greenish, in spike-like clusters......................... *E. scoparia* (Mi)
1. Fls 4-7mm, white or pink .... *E. multiflora* (Ma, Mi, I)

PRIMULACEAE

The superficial resemblance between members of this family is almost nil, but they have a lot of botanical features in common. The corolla is tubular at the base and usually 5-lobed. The stamens are inserted into the corolla tube opposite the petals, and the ovary is nearly always superior with a single cell and a simple style.

Lysimachia minoricensis, a small endemic shrub of Minorca with reddish-green axillary flowers, is omitted from the key because it is extinct in the wild state. It survives in some gardens in Minorca and elsewhere.

KEY

1. Corolla 2-lipped, upper 3 petals longer than the lower 2 .... *Coris monspeliensis* (Ma, Mi, I)
1. Corolla actinomorphic or absent
  2. Fls axillary, or in branched or simple racemose infce
    3. Corolla absent; calyx teeth petaloid, pink ..... *Glaux maritima* (?Ma)
    3. Corolla present, sometimes very small; calyx not petaloid
      4. Corolla white or greenish-white
        5. Corolla less than 2mm, shorter than calyx .... *Asterolinon linum- stellatum* (Ma, Mi, I)
        5. Corolla 2-3mm, longer than calyx .. *Samolus valerandi* (Ma, Mi, I)
      4. Corolla pink, red or blue .... *Anagallis*
  2. Fls solitary or terminal, or apparently so
    6. Fls drooping, petals reflexed........ *Cyclamen balearicum* (Ma, Mi, I)
    6. Fls erect, pale greenish-yellow, with spreading petals .... *Primula vulgaris* ssp *balearica* (Ma, endemic ssp)

Samolus valerandi Brookweed; Muraje de arrayos, Pamplina d'agua; Bunge; Mouron d'eau.
Cyclamen Sowbread; Pa de porc, Patata morena, Xiclines, Pamporcino; Alpenveilchen; Cyclamen.
Primula vulgaris Primrose; Prima vera, Herba de Sant Pau; Primel; Primevère.

ANAGALLIS

1. Fls pink
  2. Stems prostrate, rooting at nodes; fls up to 14mm diam. (boggy ground) .... *A. tenella* (Ma)
  2. Stems erect, not rooting; fls less than 1mm diam....... *A. minima* (Mi)
1. Fls red or blue
  3. Shrubby pnnl; fls up to 2cm .... *A. linifolia* (Ma, I)
  3. Annl
    4. Marginal hairs on petals absent or few (fls blue) .... *A. foemina* (Ma, Mi, I)

4. Marginal hairs on petals numerous (fls red or blue)........................ .... *A. arvensis* (Ma, Mi, I)

Anagallis arvensis Scarlet Pimpernel; Moragues, Picapoll, Morrón; Gaucheil; Morgeline.

LIMONIACEAE

The only genus here is *Limonium* (Sea-lavender; Ben rojo, Ensopegalls, Oreja de liebre, Espantazorras; Widerstoss; Lavande-de-mer). Identification is usually difficult, and hybrid swarms involving several species lead to some confusion. This key is based on descriptions in the FE and is not intended to take account of hybrids, which are very much the province of the specialist.

Plants may be annual or perennial. Leaves are normally all basal. The inflorescence is branched, and bears spikelets consisting of one or more flowers surrounded by three scale-like bracts. Of these the inner and outer bracts are larger and more important for identification purposes.

Spikelets are further aggregated into spikes, the number of spikelets per centimetre in a spike being fairly constant for one species.

At the junction between branches of the inflorescence there are small scales. The size of these is important, and the measurements here refer to the largest scales generally to be found at the lowest point where the inflorescence branches.

Non-flowering branches are characteristic of some species, and important for identification. They have only a single scale at the apex. Sterile branches are different – they have abortive spikelets with several scales at the apex. They occur in some species at the end of the summer and under unfavourable conditions, and are not of much significance for identification.

The corolla is tubular, with five spreading pink or violet lobes.

1. Flowering stem winged...................... *L. sinuatum* (Ma, introduced)
1. Flowering stem not winged
2. Bracts surrounding calyx with long fine apical awn .... *L. ferulaceum* (Ma, Mi)
2. Bracts unawned, though midrib sometimes excurrent with acute point
3. Inner bract with midrib excurrent into conspicuous acute point
4. Spikes up to 2cm long, spklts 1-flowered....... *L. tremolsii* (Ma, Mi)

4. Spikes 2cm or more; spklts 2-flowered.......................................
L. inarimense (Ma, Mi, I: ssp *ebusitanum* is endemic here; the type occurs only in Italy)

3. Midrib of inner bract not or hardly excurrent (up to 0.5mm in *L. girardianum*)

5. Spikes with less than 4 spklts per cm

6. Inner bract covered in small tubercles

7. Spklts curved.................................. *L. echioides* (Ma, Mi, I)

7. Spklts straight...................................... *L. graecum* (Ma, Mi)

6. Inner bract without tubercles

8. Spklts straight...................................... *L. graecum* (Ma, Mi)

8. Spklts curved............................. *L. duriusculum* (Ma, Mi, I)

5. Spikes with 4 or more spklts per cm

9. Lvs not more than 1cm long, puberulent (only apparent in fresh specimen) ....
L. caprariense (ssp *caprariense* Cabrera and Ma, endemic: ssp *multiflorum* Ma, Mi, I, endemic)

9. Mature lvs more than 1cm long, not puberulent

10. Inner bract 3mm long or less

11. Lvs not longer than 50mm; longest scales 6mm; corolla pale pink .... *L. delicatulum* (Ma)

11. Lvs 60mm or more; scales up to 15mm; corolla pale violet ....
.... *L. biflorum* (Ma, endemic)

10. Inner bract at least 3.5mm long, usu. more

12. Lvs often viscid; inner bract with bright red membranous margin .... *L. auriculae-ursifolium* (Ma, Mi)

12. Lvs not viscid; inner bract without bright red margin

13. Lvs revolute

14. Lvs with 1 basal vein; infce lax, 4-6 spklts per cm ....
.... *L. minutiflorum* ssp *balearicum* (Ma, endemic)

14. Lvs with 3 basal veins; spklts contiguous ....
...... *L. majoricum* (Ma, endemic, but possibly now extinct)

13. Lvs flat

15. Plant with many non-flowering branches; spikes with about 4 spklts per cm ....
L. oleifolium (ssp *oleifolium* Ma, Mi, I: ssp *pseudodictyocladum* Ma, endemic)

15. Non-flowering branches absent; spikes with 6 or more spklts per cm

16. Plant more than 20cm high; bracts often lf-like ....
.... *L. ramosissimum* (Ma)

16. Plant less than 20cm; bracts never lf-like
17. Lvs blunt; spikes with 7-9 spklts per cm .... *L. gougetianum* (Ma, Mi, I)
17. Lvs pointed; spikes with 10-15 spklts per cm.................... .... *L. girardianum* (I)

OLEACEAE

This is the family of *Olea europaea* (Olive; Ullastro, Olivera, Aceituno; Olbaum; Olivier), one of the most ancient crops of the Mediterranean. A psalm written thousands of years before the invention of sun-tan lotions celebrates the use of olive-oil as a skin salve: 'wine that maketh glad the heart of man; and oil to make him a cheerful countenance', and olive oil is still an important product of Majorca, though now mainly used for food.

In Homeric times olive-oil seems to have been a luxury for the toilet of the wealthy, but Homer makes no mention of the cultivation of the tree, which probably originated further east than Greece. It gradually spread westward, eventually via the Romans (whose prescriptions for a healthy life included 'wine within and oil without'), to the Roman province in Spain and the Balearic Islands.

The family includes shrubs, trees and climbing perennials. The leaves are entire or pinnate. The calyx is generally tubular and 4-lobed, and the corolla, when present, is also 4-lobed, sometimes with free petals. The ovary is superior, with two cells, and the style is simple with a capitate or bifid stigma. There are usually two stamens.

KEY

1. Tree, up to 25m, with pinnate lvs; fr. dry, winged .... *Fraxinus angustifolia* (Ma)
1. Small tree or shrub with simple lvs
2. Plant spiny; fls white; fr. green .... *Olea europaea* (Ma, Mi, I, introduced*)
2. Plant without spines; fls greenish-yellow; fr. bluish-black .... *Phillyrea*

PHILLYREA

Lvs dimorphic, adult lvs narrow elliptical, 1-6cm, entire or toothed, with 7-11 pairs of veins.. *P. latifolia* (Ma)

* This refers to the wild olive, ssp *sylvestris*. The cultivated tree is not spiny.

Lvs all alike, linear to lance-shaped, with 4-6 pairs of (indistinct) veins...... .... *P. angustifolia* (Ma, Mi, I)

GENTIANACEAE

There are no gentians here, and the colours are pink and yellow, not blue. The plants here are hairless, with opposite, entire leaves, each pair usually at right angles to the pair above and below. The flowers are actinomorphic, usually arranged in a dichasial cyme.

KEY

1. Corolla lobes 4 or 5; cauline lvs not connate in pairs
2. Calyx deeply divided into keeled, linear lobes; corolla lobes usu. 5 (sometimes 4 in small plants).................................... *Centaurium*
2. Calyx bell-shaped, lobes short, triangular ..... *Cicendia filiformis* (Mi)
1. Corolla lobes 6-8; cauline lvs usu. connate in pairs........................... .... *Blackstonia perfoliata* (Ma, Mi, I)

CENTAURIUM

1. Corolla yellow, sometimes pinkish yellow...... *C. maritimum* (Ma, Mi)
1. Corolla pink
2. Fls in spike-like infce................................ *C. spicata* (Ma, Mi, I)
2. Fls not in spike-like infce
3. Rosette of basal lvs present
4. Rosette lvs 0.8-2cm wide .... *C. erythraea* (Ma, Mi)
4. Rosette lvs about 0.3cm wide .... *C. linariifolium* (?Ma)
3. Rosette absent at flowering time, or with only few withering lvs (corolla commonly deep carmine)...................... *C. pulchellum* (Ma)

C. enclusense (Mi) is described in the FE as a possible ssp of *C. erythraea*, divaricately branched from base and with a large flower.

APOCYNACEAE

This is a family found mainly in the tropics and subtropics, with only two species here. Both have actinomorphic, 5-petalled flowers, with the petals united into a tube at the base. In both species the 8 petals are unsymmetrical in shape, producing a Catherine wheel effect.

KEY

Procumbent, trailing plant; fls solitary, blue, white or very pale pink .... *Vinca difformis* (Ma, Mi, I)

Erect shrub; fls clustered, deep pink.. .... *Nerium oleander* (I) (Ma, introduced)

Nerium oleander Oleander; Rosa reial, Rosa laurel; Oleander; Laurier rose.
Vinca difformis Periwinkle; Proenga, Vincapervinca, Hierba lechera; Immergrün; Pervenche.

ASCLEPIADACEAE

These flowers are rather small and dull-coloured, but intriguingly different, and well worth a close look.

The flower is actinomorphic, with five petals and a corona – a ring of scales fused either to the inside of the corolla tube or to the ring of stamens. The corona may be five similar scales, or ten scales with two different types alternating. The stigma is single and disc-like, though arising from two styles, and there are five stamens, fused into a tube, often with the anthers fused into a circle above the stigma. The whole effect is rather geometrical and elegant.

KEY

1. Shrubby erect pnnl, 1m or more .... *Gomphocarpus fruticosus* (Ma, I, introduced)
1. Plant under 1m, or climbing
   2. Climbing plant; fls white or pink .... *Cynanchum acutum* (Ma)
   2. Plant erect or weakly climbing; fls green or blackish-purple.............. .... *Vincetoxicum*

VINCETOXICUM

Fls greenish-white .... *V. hirundinaria* (Ma, Mi)
Fls purplish-black.. *V. nigrum* (Ma, Mi)

RUBIACEAE

Probably the best known member of this family is *Galium aparine* (Goosegrass, Cleavers; Herba de gallina, Amor d'hortolà, Arpegalós, Làrgulo;

Klebriges Labkraut; Gratteron), spreading its small green fruits covered with tiny hooked spines by clinging to the clothes of passers-by almost anywhere in Europe. The plant is typical of most of the family, with its weak, straggling stems, whorls of leaves, and tiny flowers usually in branched axillary inflorescences.

KEY

1. Fls solitary, sessile
  2. Plant rigid; fls in dense terminal infce with 4 rows of imbricate bracts (usually more noticeable than the tiny fls) .... *Crucianella*
  2. Plant straggling; fls axillary .... *Valantia*
1. Fls in branched terminal or axillary infce
  3. Corolla tube longer than lobes
    4. Lvs in whorls of 4, or paired .... *Asperula*
    4. Lvs in whorls of more than 4
      5. Fls bluish-lilac.. *Asperula*
      5. Fls not bluish-lilac (rarely pinkish, never bluish)
        6. Fls in terminal heads; calyx with 4-5 persistent obvious teeth .... *Sherardia arvensis* (Ma, Mi, I)
        6. Fls not in terminal heads; calyx absent, or with very small teeth..... .... *Galium*
  3. Corolla tube shorter than lobes
    7. Corolla 5-lobed; fr. fleshy .... *Rubia peregrina* (Ma, Mi, I; ssp *peregrina*, and endemic ssp *angustifolia*)
    7. Corolla usu. 4-lobed; fr. dry .... *Galium*

CRUCIANELLA

1. Pnnl; lvs in whorls of 4, leathery .... *C. maritima* (Ma, Mi, I)
1. Annl; lvs in whorls of more than 4, not leathery
  2. Bracts narrowly lanceolate, free.................. *C. angustifolia* (Ma, I)
  2. Bracts broadly lanceolate, often connate at base... *C. latifolia* (Ma, I)

ASPERULA

1. Fls bluish-lilac; at least some lvs in whorls of more than 4 .... *Asperula arvensis* (Ma, Mi, I)
1. Fls not bluish-lilac; lvs in whorls of 4 or less
  2. Lvs mostly in pairs .... '*A. paui*' (I, probably a ssp of *A. aristata*)
  2. Lvs in whorls of 4
    3. Lvs with short point or none; fls white .... *A. laevigata* (Ma, Mi)

3. Lvs with fine point; fls white inside, pink outside .... *A. cynanchica* (Ma, I)

GALIUM

1. Annl
 2. Fr. more than 2mm; corolla white or creamish
  3. Lf margin with bristles pointing towards base
   4. Fr. covered in hooked bristles .... *G. aparine* (Ma, Mi, I)
   4. Fr. warty, without bristles .... *G. tricornutum* (Ma, Mi, I)
  3. Lf margin with bristles pointing towards apex .... *G. verrucosum* (Ma, Mi, I)
 2. Fr. less than 2mm; corolla yellowish, red or purple
  5. Fls axillary, solitary or paired; fr. cylindriacal, with hooked bristles towards apex .... *G. murale* (Ma, Mi, I)
  5. Fls in many-flowered axillary racemes; fr. globular
   6. Corolla purple .... *G. setaceum* (Ma)
   6. Corolla greenish, yellow or red
    7. Stems nearly smooth above; corolla yellowish-red .... *G. divaricatum* (Ma)
    7. Stems 4-angled, angles with downwardly directed prickles; corolla greenish inside, reddish outside .... *G. parisiense* (Ma, Mi)

1. Pnnl
 8. Corolla less than 3mm, bright purple (mountain scrub) .... *G. balearicum* (Ma, endemic)
 8. Corolla 3mm or more, not bright purple
  10. Fr. globose.. *G. palustre* (Ma, Mi, I)
  10. Fr. egg-shaped
   11. Longest lvs more than 3mm wide, often curved .... *G. crespianum* (Ma, I, endemic)
   11. Longest lvs less than 3mm wide, usu. straight
    12. Lvs glaucous... *G. cinereum* (Ma)
    12. Lvs green
     13. Lvs 5-11mm, midrib prominent, more than half width lf .... *G. corrudifolium* (Ma, I)
     13. Lvs 10-30mm, midrib slender, less than half width lf .... *G. lucidum* (Ma, I)

VALANTIA

Plant rough hairy; peduncle without dorsal horn .... *V. hispida* (Ma, I)
Plant slightly hairy; peduncle with dorsal horn ...... *V. muralis* (Ma, Mi, I)

CONVOLVULACEAE

Many of the plants in this family are climbers; some are shrubs and herbaceous plants. They generally have simple alternate leaves, and a funnel-shaped corolla which is sometimes lobed. The five segments of the perianth are easy to recognize even when the petals are fused and the corolla almost unlobed.

For most people the most familiar species will be *Convolvulus arvensis* (Cornbine, Bindweed; Marañuela, Esquiluntza, Campaneta; Ackerwinde; Liseron)

KEY

1. Greyish shrublet; fls 3mm, pink or yellow, in terminal clusters; corolla lobes longer than tubular part .... *Cressa cretica* (Ma, I)
1. Plant not as above; corolla lobes equalling tubular or funnel-shaped or shorter, or corolla without lobes*
 2. Parasitic plant without green parts; stem thread-like, twining; fls up to 12mm in globular clusters.. *Cuscuta*
 2. Plant not parasitic, leaves green or greyish-green; fls solitary or more than 12mm
  3. Bracteoles large, more or lesss concealing sepals........... *Calystegia*
  3. Bracteoles not or hardly concealing sepals
   4. Stigmas 2, linear... *Convolvulus*
   4. Stigma 1, club-shaped.. *Ipomoea*

CUSCUTA Dodder; Barbes de caputxi, Cabells d'àngel; Seide; Cuscute.

1. Calyx tube golden yellow, shiny and reticulate when dry ....
 .... *C. approximata* (Ma)
1. Not as above
 2. Fls usu. 3-4mm in globular clusters 7-10mm diam...........................
 .... *C. epithymum* (Ma, Mi, I)
 2. Fls 1.5-3mm
  3. Clusters 4-6mm diam. .... *C. planiflora* (Ma)
  3. Clusters 2-3mm diam. .... *C. campestris* (Ma)

* Flowers with the corolla divided almost to the base may be found as an occasional oddity here. This seems to be particularly common in *Convolvulus cantabricus*, and in wet weather.

CALYSTEGIA

Lvs kidney-shaped; fls pink (maritime) .... *C. soldanella* (Ma, Mi, I)
Lvs heart or arrow-shaped; fls usu. white............. *C. sepium* (Ma, Mi, I)

CONVOLVULUS

1. Lvs sessile, or attenuated below, with no distinct stalk
 2. Corolla very pale blue centre yellow .... *C. pentapetaloides* (Ma, I)
 2. Corolla pink
  3. Plant with appressed silky hairs; peduncles all much shorter than subtending bracts...................................... *C. lineatus* (Ma, I)
  3. Plant with spreading hairs at least below; at least lower peduncles longer than subtending bracts .... *C. cantabrica* (Ma, Mi)
1. Lvs with distinct stalk
 4. At least upper lvs distinctly lobed or divided................................ .... *C. althaeoides* (Ma, Mi, I)
 4. Lvs more or less entire, often hastate or sagittate at base
  5. Sepals blunt to notched .... *C. arvensis* (Ma, Mi, I)
  5. Sepals acute
   6. Corolla 18-30mm.. *C. valentinus* (I)
   6. Corolla 7-12mm .... *C. siculus* (Ma, Mi, I)

IPOMOEA

Fls white or pale yellow, sometimes with purple centre......................... .... *I. stolonifera* (? Is Cabrera)
Fls pink to purple .... *I. sagittata* (Ma, Mi, I)

BORAGINACEAE

This family is named after *Borago* (the only species here *Borago officinalis* Borage; Borraja, Borraina, Orella d'os, Pa i peixet; Borretsch; Bourrache), a large hispid annual with five-petalled gentian-blue flowers.

The petals are united into a tube (sometimes very short) at the base.

The buds are arranged in cymes with the axis coiled up; the coil unwinds as the flowers open, lowest first, so that the proximal part of the cyme with the developing fruits is straight. This is called a 'scorpioidal cyme', and is very characteristic of this family – as are the bright blue flowers, though neither character is invariable.

The fruit in this family is usually two or four nutlets.

KEY

1. Fls tubular, drooping; corolla usu. yellow, at least peripherally; lvs glaucous, hairless, clasping stem lvs; nutlets connate in 2 pairs .... *Cerinthe minor* (Mi, introduced)
1. Fls and lvs various, not as above; nutlets not connate in two pairs
 2. Corolla yellow, yellowish or greenish-white, or white, not pinkish or bluish-white
  3. Infce drooping, subtended by pair of large, hairy, elliptical lvs; corolla lobes much shorter than tube .... *Symphytum tuberosum* (Ma, Mi)
  3. Fls usu. spreading or erect; lvs not as above; corolla lobes at least 1/3 length tube
   4. Fls bright yellow
    5. Calyx 3.5-4mm; corolla lobes softly glandular hairy on both surfaces .... *Neatostema apulum* (Ma, I)
    5. Calyx 5-6mm; corolla hairless on outside .... *Alkanna lutea* (Ma, Mi, I)
   4. Fls white or cream coloured (sometimes with yellow centre)
    6. Style terminal.. *Heliotropium*
    6. Style arising from between lobes of ovary
     7. Corolla 3-6mm, yellowish or greenish-white .... *Lithospermum officinale* (Ma)
     7. Corolla 6-9.5mm, white, rarely purplish .... *Buglossoides arvensis* (Ma, Mi, I)
 2. Fls blue, purple, pink, or red, or pinkish or bluish-white
  8. Corolla lobes pointed...................... *Borago officinalis* (Ma, Mi, I)
  8. Corolla lobes rounded
   9. Plant with bristly hairs
    10. Corolla with 5 longitudinal bands of hairs inside .... *Buglossoides arvensis* (Ma, Mi, I)
    10. Corolla not as above,though sometimes hairy inside
     11. Corolla mouth transverse, tube cylindrical, straight, or curved ... . *Anchusa*
     11. Corolla mouth usu. oblique, tube usu. much wider at mouth than base .... *Echium*
   9. Plant without bristly hairs
    12. Fls 4mm or less, pale blue
     13. Lf-like bracts in infce; nutlets with bristly spines .... *Lappula squarrosa* (?Ma)
     13. Infce without bracts; nutlets not spiny .... *Myosotis arvensis* (Is Formentera)

12. Fls 7-9mm, pale blue with conspicuous darker veins, or red .... *Cynoglossum*

Symphytum tuberosum Comfrey; Sinfito menor, Consoldo or Consueldo menor; Knolliger Beinwell; Consoude.
Lithospermum officinale Grummel, Gromwell; Aljófar, Mill bord, Mijo de sol; Echter Steinsame; Grémil.

HELIOTROPIUM

1. Lvs fleshy, without hairs .... *H. currassavicum* (Ma, Mi, I)
1. Lvs hairy, not fleshy
   2. Calyx less than 1/4 lobed, almost concealing fr.; fr. with one seed .... *H. supinum* (Mi)
   2. Calyx lobed to base, not concealing fr.; fr. with 4 nutlets .... *H. europaeum* (Ma, Mi, I)

H. europaeum Heliotrope; Gira-sol, Tornasol, Herba berruguera; Sonnenwende; Herbe aux verrugues, Tournesol.

ECHIUM Bugloss; Bolenga, Viborera; Natternkopf; Vipérin.

1. All stamens completely inside corolla tube
   2. Calyx with appressed hairs, up to 15mm in fr., lobes 3-6mm wide at base.. *E. parviflorum* (Ma, Mi, I)
   2. Calyx with spreading hairs, up to 10mm in fr., lobes 2-3mm wide at base.. *E. arenarium* (Ma, I)
1. At least 1 or 2 stamens protruding from corolla tube
   3. Fls flesh-pink or pale pinkish or bluish-white, brown or blue-grey when dry
      4. Corolla 13-18mm, flesh pink............... *E. asperrimum* (Ma, Mi, I)
      4. Corolla 10-12mm, pale pinkish or bluish-white .... *E. italicum* (Ma, Mi, I)
   3. Fls blue, purple or carmine, reddish-blue or purple when dry
      5. Corolla hairy on veins and margins only .... *E. plantagineum* (Ma, Mi, I)
      5. Corolla more or less uniformly hairy .... *E. sabulicola* (Ma, Mi, I)

ANCHUSA

Local names usually as *Echium* Annl; corolla lobes unequal .... *A. arvensis* (?Ma)
Bnnl or pnnl; corolla lobes equal .... *A. azurea* (Ma, Mi, I)

PLATE 4. Actinomorphic flowers with more than 10 petals. Compound flowers, the florets actinomorphic or zygomorphic.

	Key page		Key page
Aizoaceae			
1. *Carpobrotus acinaciformis*	41	19. *Anthemis maritima*	148
2. *Mesembryanthemum crystallinum*	42	20. *Achillea ageratum*	142
Ranunculaceae		21. *Chrysanthemum coronarium*	148
3. *Ranunculus ficaria*	50	22. *'Senecio rodriguezii'*	148
Globulariaceae		23. *Calendula arvensis*	142
4. *Globularia alypum*	131	24. *Carlina corymbosa*	149
Dipsacaceae		25. *Cirsium arvense*	149
5. *Cephalaria squamiflora* ssp *balearica*	137	26. *Galactites tomentosa*	141
6. *Scabiosa cretica*	137	27. *Crupina crupinastrum*	150
7. *S. atropurpurea*	137	28. *Centaurea aspera*	150
Compositae		29. *Crupina crupinastrum*	150
8. *Eupatorium cannabinum*	141	30. *Carthamnus lanatus*	140
9. *Bellis annua*	145	31. *Cichorium intybus*	145
10. *Bellium bellidioides*	143	32. *Hyoseris scabra*	152
11. *Aster tripolium*	146	33. *Urospermum dalechampii*	152
12. *Conyza canadensis*	146	34. *Hypochaeris achryophorus*	152
13. *Helichrysum ambiguum*	146	35. *Tragopogon hybridus*	152
14. *Phagnalon sordidum*	147	36. *Reichardia tingitana*	152
15. *P. saxatile*	147	37. *Launaea cervicornis*	144
16. *Pallenis spinosa*	142	38. *Aetheorhiza bulbosa* ssp *willkommii*	144
17. *Asteriscus maritimus*	147	39. *Lactuca tenerrima*	153
18. *Santolina chamaecyparissus*	140		

1
2
19
27
3
6
4
24
39
9
35
20
7
10
13
30
18
15
14
17
23
29
16
21
28
25
26
33
22
34
38
36
12
32
11
37
5
8
31

CYNOGLOSSUM Hound's tongue; Llapassa, Viniebla, Lengua de perro; Hundszunge; Langue-de-chien.

Fls pale blue with darker veins; lvs hairy, green .... *C. creticum* (Ma, Mi, I)

Fls reddish purple; lvs grey or white with dense hairs .... *C. cheirifolium* (Ma, Mi, I

VERBENACEAE

The few species of this family which occur wild here vary from small herbs to a largish shrub. The flowers are small, five-petalled and two-lipped (often only slightly so), coloured pinkish to violet or white.

KEY

1. Shrub 1-6m; lvs palmate; fls in terminal cluster .... *Vitex agnus-castus* (Ma, Mi, I)
1. Herb, sometimes woody at base, up to 100cm, usu. less
   2. Lvs pinnately lobed or pinnatisect .... *Verbena*
   2. Lvs simple .... *Lippia*

Vitex agnus-castus Chastity bush; Alo, Arbre de San Josep, Eliasiera; Mönchspeffer; Gattilier, supposed by the ancients to dull the pangs of love. Its generic name is derived from the Latin *vieo*, to bend, twist together, or plait, and its supple twigs are still used in basket making.

VERBENA

Pnnl; usu. erect; stems rough on the angles .... *V. officinalis* (Ma, Mi, I)

Annl; usu. procumbent; stems not rough .... *V.supina* (Mi)

LIPPIA

Fl. spikes 5-7mm diam.................................. *L. nodiflora* (Ma, Mi, I)

Fl. spikes 9-12mm diam. .... *L. canescens* (Ma)

CALLITRICHACEAE

The only genus is *Callitriche*, typically aquatic plants with weak submerged stems and entire opposite leaves, the terminal ones generally forming a floating rosette. They may also be found on mud. For identification ripe fruit and a hand lens are needed. The key here only identifies typical fertile aquatic forms. Sterile forms or terrestrial forms are often very different.

CALLITRICHE

1. Plant entirely submerged; fr. wider than long, separating into unwinged or narrowly winged seeds .... *C. truncata* (Mi)
1. Plant with floating rosette of lvs; seeds broadly winged
  2. Rosette lvs broadly elliptical to almost circular narrowed abruptly into the stalk; styles becoming recurved, rarely persistent in fr. .... *C. stagnalis* (Ma, Mi)
  2. Rosette lvs elliptical or slightly obovate; styles deflexed, appressed to fr. .... *C. brutia* (Ma, Mi)

LABIATAE

This is a large and varied family, including many well known culinary and aromatic herbs. The leaves are in opposite pairs, each pair usually at right angles to the one above and the one below it. The leaves are usually smaller in the upper part of the stem, and often merge gradually into the bracts in the inflorescence.

The flowers are variously arranged in the leaf axils. As the pedicels and peduncles if present are usually short they appear to be in whorls, and the term whorl is used here even if not always exact. The calyx is tubular, often two lipped with three teeth above and two below, and the corolla likewise only with the three lobes below. The ovary is superior with a single style.

KEY

1. Upper lip of corolla absent .... *Teucrium*
1. Upper lip of corolla present, sometimes very small
  2. Bracts markedly different in shape and often colour, from lvs
    3. Stamens 2 .... *Salvia*
    3. Stamens 4
      4. Plant entirely herbaceous, or rarely slightly woody at base
        5. Fls cream or violet in dense terminal head .... *Prunella*
        5. Fls white or pink in loose terminal panicle .... *Origanum*
      4. Shrub .... *Lavandula*
  2. At least lower bracts lf-like; bracts usu. reduced progressively in size, ornament or petiole length towards tip of infce

6. Stamens 2 .... *Rosmarinus officinalis* (Ma, Mi, I)
6. Stamens 4
7. Calyx with erect dorsal scale behind upper lip .... *Scutellaria balearica* (Ma, endemic)
7. Calyx not as above
8. Calyx with 13 veins or more
9. Calyx with 15 veins or more
10. Calyx funnel-shaped, greatly enlarged, spiny .... *Moluccella laevis* (Ma)
10. Calyx not as above
11. Calyx with 20-22 veins .... *Thymus*
11. Calyx with 15 veins .... *Nepeta*
9. Calyx with 13 veins
12. Corolla tube curved, dilated above middle .... *Melissa officinalis* (Ma, Mi, I)
12. Corolla tube straight, not dilated
13. Whorls of more than 20 fls .... *Mentha*
13. Whorls of 20 fls or less, usu. many less
14. Lvs entire,
15. Stamens shorter than corolla, curved .... *Micromeria*
15. Stamens protruding from corolla, straight .... *Thymus*
14. Some lvs toothed or serrated
16. Calyx tube curved, middle constricted, swollen at base .... *Acinos arvensis* (Ma)
16. Calyx tube straight, not constricted or swollen .... *Calamintha*
8. Calyx with less than 13 veins
17. Corolla 10mm or more
18. Lvs shiny, hairless or nearly so .... *Prasium majus* (Ma, Mi)
18. Lvs more or less hairy
19. Lvs with star-shaped hairs .... *Phlomis italica*(Ma, Mi, endemic)
19. Lvs without star-shaped hairs
20. Lower lip corolla with lateral lobes minute or 0 .... *Lamium amplexicaule* (Ma, Mi, I)
20. Lower lip of corolla with distinct lateral lobes
21. Calyx lobes mucronate or awned, often toothed............... .... *Ballota*
21. Calyx lobes without mucro, awn, or teeth .... *Stachys*
17. Corolla less than 10mm
22. Fls yellow .... *Ajuga*

22. Fls white, pink or purplish
23. Whorls crowded into terminal heads
24. Corolla with 4 almost equal lobes .... *Mentha*
24. Upper lip corolla entire to bifid, lower 3-lobed ....
.... *Thymus*
23. Whorls not crowded into terminal heads
25. Annl
26. Upper lobe calyx much larger than other 4......................
.... *Sideritis romana* (Ma, Mi, I)
26. Calyx teeth equal or slightly unequal .... *Stachys*
25. Pnnl
27. Lvs densely tomentose, at least below...........................
.... *Marrubium vulgare* (Ma, Mi, I)
27. Lvs hairy, not densely tomentose....................... *Mentha*

Rosmarinus officinalis Rosemary; Romaní, Romero; Rosmarin; Romerin.
Melissa officinalis Balm; Herba cidrera, Hierba cidra; Melisse; Mélisse.
Acinos arvensis Basil-thyme; Albahaca, Alfábrega; -;-.
Phlomis italica (which is endemic here and does not occur in Italy), is called Ble de Frare, Estepa blanera or Candilera locally.

AJUGA Ground pine; Iveta, Iva artètica; Günsel; Bugle petit-pin.
Lf segments 0.5-2mm wide .... *A. chamaepitys* (Ma)
Lf segments 3-8mm wide.. *A. iva* (Ma, Mi, I)

TEUCRIUM
1. Spiny hedgehog plant................ *T. subspinosum* (Ma, Mi, endemic)
1. Plant not spiny
2. Corolla yellow .... *T. flavum* (Ma, I)
2. Corolla red, pink, purplish or white
3. Lvs 1 or 2 x pinnatisect
4. Annl; fls 15-20mm .... *T. botrys*(Ma)
4. pnnl; fls ca 5mm .... *T. campanulatum* (Ma)
3. Lvs simple
5. Fls in terminal heads
6. Lvs entire, or crenate in distal 1/4; calyx 5-6mm ....
.... *T. cossonii* (Ma, I, endemic)
6. Lvs crenate in at least distal 1/2; calyx 2.5-5mm........................
.... *T. polium* (Ma, Mi, I, with endemic ssp see below)
5. Fls not in terminal heads

7. Calyx teeth with terminal bristle ... *T. asiaticum* (Ma, Mi, endemic)
7. Calyx teeth without terminal bristle
 8. Stoloniferous pnnl .... *T. scordium* (Ma, Mi)
 8. Small shrub, not stoloniferous
  9. Lvs usu. entire, grey-tomentose beneath; calyx teeth 1/3 x tube .... *T. marum* (Ma)
  9. Lvs usu. deeply serrate and hairy beneath; calyx teeth 2/3 x tube.. *T. chamaedrys* (Ma, Mi)

T. polium. Two subspecies occur here. *T. polium* ssp *polium* is here a maritime shrublet, grey woolly with small white flowers or rather less woolly with rosy pink flowers. *T. polium* ssp *pii=fontii*, with slender flowering stems, and axillary fascicles is an inland endemic plant.

T. chamaedrys Wall germander; Alzineta, Camedris, Carasquilla; Echter Gamander; Germandrée, Petit chêne.

BALLOTA

Calyx irregularly toothed, with up to 10 lobes with very small mucro .... *B. hirsuta* (Ma, I)
Calyx with 5 regular awned or mucronate lobes.......... *B. nigra* (Ma, Mi)

Ballota nigra Black horehound; Malrubí bord, Balota; Stinkandorn; Marrube noire, Ballote.

STACHYS

1. Densely tomentose pnnl .... *S. germanica* (Ma)
1. Hairy annl
 2. Fls yellow or yellow and white .... *S. ochymastrum* (Ma, Mi, I)
 2. Fls reddish-purple to pink or white
  3. Whorls 6-12; calyx teeth about equalling tube....... *S. arvensis* (Ma)
  3. Whorls usu. more than 12; calyx teeth unequal, upper 2 about equalling tube .... *S. brachyclada* (I)

NEPETA Catmint; Nepte, Herba del gat; Katzenminze; Herbe aux chats

Corolla 7-10mm, white with purple spots...................... *N. cataria* (Mi)
Corolla 10-12mm, white, pink or bluish-violet .... *N. nepetella* (Mi)

PRUNELLA

Upper lvs entire or toothed: fls violet.......................... *P. vulgaris* (Mi)
Upper lvs pinnatifid: fls cream .... *P. laciniata* (Ma)

Prunella laciniata Self-heal; Herba de la princesa, Hierba de las heridas; Braunelle; Brunelle.

CALAMINTHA Calamint; Borriol; Nevada; Kalaminthe; Sarriette.
Corolla pink or lilac with white spots on lower lip ....
.... *C. sylvatica* (Ma, Mi, I) (ssp *ascendans* here)
Corolla white or lilac................. *C. nepeta* (Ma, I) (ssp *glandulosa* here)

MICROMERIA Local names: Picardia, Mirra, Tem, Ajedrea.
1. Calyx with dense long spreading hairs.................. *M. nervosa* (Ma, I)
1. Calyx with short hairs or none
  2. Lvs with revolute margins
    3. Lvs 5-12mm... *M. graeca* (Ma, I)
    3. Lvs 2-3mm .... *M. inodora* (Ma, I)
  2. Lvs flat.. *M. filiformis* (Ma, Mi, I)

ORIGANUM Marjoram; Orenga, Orégano; Dost; Origan.
Bracts twice length calyx, gland dotted, hairless; corolla white ....
.... *O. virens* (Ma)
Bracts slightly longer than calyx, shaggy haired; corolla pink .... *O. majoricum* (Ma, not recently, ?hybrid origin, *O. virens x O. majorana*, the last a much cultivated pot herb)

THYMUS Thyme; Sajolida, Farigola, Timó; Thymian; Thym.
1. Calyx tube dorsally flattened, veins 20-22 ..... *Thymus capitatus* (Ma, I)
1. Calyx tube more or less bell-shaped, with 10-13 veins
  2. Calyx 3-4mm; corolla whitish to pale purple ....
  .... *T. vulgaris* (I, and cultivated Ma)
  2. Calyx 4.5-6mm; corolla purple....................... *T. richardii* (Ma, I)*

MENTHA Mint; Menta; Minze; Menthe.
1. Calyx weakly 2-lipped, upper teeth shorter and wider than lower........
.... *M. pulegium* (Ma, Mi)
1. Calyx teeth more or less equal
  2. Lvs sessile or very short stalked, except sometimes lowest pair
    3. Lvs lanceolate to narrowly ovate

* *T. richardii* has three subspecies: ssp *richardii* occurs only in Ma and Jugoslavia, ssp *nitidus* is endemic in an island off the West coast of Sicily, ssp *ebusitanus* is endemic in Ibiza.

4. Lvs or nearly so on both sides .... *M. spicata* (Ma)
4. Lvs tomentose below, usu. hairy above .... *M. longifolia* (Ma)
3. Lvs ovate to almost round, usu. hairy, grey or white with dense cottony hairs beneath .... *M. suavolens* (Ma, Mi, I)
2. Lvs obviously stalked
5. Infce of 2-3 congested whorls forming head up to 20mm diam. (sometimes with 1-3 distant whorls beneath); pedicels hairy .... *M. aquatica* (Ma, Mi, I)
5. Infce usu. of many congested whorls of 12-18mm diam., often widely spaced below; pedicels usu. hairless.............................. .... *M. piperita* (Ma, introduced)

LAVANDULA Lavender; Gallanda, Espígol, Lavanda; Lavandel; Lavande.
1. Lvs pinnately toothed or pinnatifid .... *L. dentata* (Ma, I)
1. Lvs entire
2. Uppermost bracts forming a crest to the infce ...... *L. stoechas* (Mi, I)
2. Uppermost bracts similar to lower, not forming a crest.................... .... *L. latifolia* (Ma, probably introduced)

SALVIA
1. Lvs pinnatifid, green.............................. *S. verbenaca* (Ma, Mi, I)
1. Lvs simple, usu. glaucous
2. Bracts exceeding corolla; calyx with spiny teeth............................ .... *S. sclarea* (Ma)
2. Bracts not exceeding calyx; calyx without spiny teeth...................... .... *S. officinalis* (Ma, Mi, I, introduced)

SOLANACEAE

There are two very important crops in this family, *Solanum tuberosum* (Potato; Pataca, Patata; Kartoffel; Pomme-de-terre), and *Lycopersicon esculentum* (Tomato; Tomaquera Tomatera; Tomate; Tomate). The mauve, white or yellow flowers of these, with five reflexed petals and anthers connivent to form a cone in the centre of the flower, are very characteristic of the family – so are the fleshy berries.

Other cultivated species are *Solanum melongena* (Egg-plant; Berenjena; Aubergine; Aubergine), *Capsicum annuum* (Sweet pepper; Picudillos, Pimiento; Spanischer Pfeffer; Piment doux, Poivrons) and *Capsicum frutescens* (Chilli; Pimiento; Chilli; Piment). Though originally from India and tropical America, these are now all characteristic ingredients of local cooking. Many ornamental species are cultivated too, especially Petunia hybrids. Not all conform to the same pattern, and tubular flowers, non-connivent anthers, and spreading stamens are all found within the family.

KEY

1. Fls yellow or yellowish-green
  2. Corolla long tubular or funnel-shaped, lobes short .... *Nicotiana*
  2. Corolla lobes almost equalling bell-shaped tube, or tube very short
    3. Anthers forming cone at centre of fl.
      4. Fls yellow, fr. a tomato .... *Lycopersicon esculentum* (introduced Ma)
      4. Fls greenish-yellow; fr. 7-8mm, red or green .... *Withania*
    3. Anthers free, not forming cone at centre of flower........................ .... *Hyoscyamus albus* (Ma, Mi, I)
1. Fls mauve, purple, pink or white
  5. Lvs in basal rosette; fls almost or completely stalkless .... *Mandragora autumnalis* (Ma, no recent record)
  5. Lvs not in basal rosette; fls in variously branched infce
    6. Annl herb or scrambling pnnl
      7. Fls large (6-19cm), funnel shaped, white; stamens included........... .... *Datura*
      7. Fls up to 1cm, not funnel-shaped, purple or white; stamens prominent .... *Solanum*
    6. Spiny shrub
      8. Corolla 25-40mm; fr. 20-30mm, yellow (looks like a tomato, but very poisonous) .... *Solanum*
      8. Corolla 11-18mm; fr. less than 10mm, reddish, orange-red or black .... *Lycium*

LYCIUM (called Cambronera, Arn blanc, Arçot de tanques locally, Lyciet in French).

1. Corolla 9-13mm; stamens usu. protruding from corolla; fr. red
  2. Leaves up to 1cm wide; corolla 11-13mm ..... *L. europaeum* (Ma, Mi)
  2. Leaves at least 1cm wide: corolla 9mm .... *L. barbarum* (Mi)
1. Corolla 13-18mm; stamens not protruding; fr. orange-red or black....... .... *L. intricatum* (Ma)

WITHANIA (called Beleño macho or Orval here)

Fls in clusters of 4-6; pedicels up to 5mm, spreading .... *W. somnifera* (Ma, Mi, I)

Fls solitary, rarely in clusters of 2-3; pedicels 10-15mm, drooping............ .... *W. frutescens* (Ma, I)

SOLANUM

1. Spiny shrub; fr. 20-30mm, yellow.............. *S. sodomeum* (Ma, Mi, I)
1. Climbing pnnl, or decumbent to erect annl; fr. finally red or black
  2. Climbing pnnl; fls usu. dark purple............... *S. dulcamara* (Ma, Mi)
  2. Decumbent to erect annl; fls white
    3. Ripe fr. black, wider than long.................. *S. nigrum* (Ma, Mi, I)
    3. Ripe fr. red, orange or yellow, usu. longer than wide ...... *S. luteum* (Ma, Mi, I)

S. dulcamara Bittersweet, Woody Nightshade; Emborrachadora, Dulciamarga; Bittersüß; Douce-amère.
S. nigrum Black nightshade; Morella vera, Schwartzer Nachtshatten; Morelle noire.
S. sodomeum is called Tomatillos del diabolo or Metzines de pometa here.

DATURA

Fls 6-8cm; fr. erect .... *D. stramonium* (I)
Fls 11-19cm; fr. pendulous .... *D. innoxia* (I)

NICOTIANA Tobacco; Tabac; Tabak; Tabac.

1. Shrub or small tree; lvs hairless, glaucous ... *N. glauca* (introduced Ma)
1. Herb; lvs sticky and hairy, not glaucous
  2. Corolla 12-17mm... *N. rustica* (I)
  2. Corolla 30-55mm................ *N. tabacum* (Escape from cultiv. ?Ma)

SCROPHULARIACEAE

This family is named after *Scrophularia* (Figwort, Betony; Escrofularia; Braunwurz; Scrofulaire), so called because it was believed to cure scrofula, a form of tuberculosis of the neck glands. From the 5th until the 18th century the disease was believed to be cured by the Royal touch, but those not fortunate enough to have access to the king made do with the plant, whose medicinal virtues included other diseases too.

The corolla is tubular, usually but not always five lobed, and the calyx four or five lobed and often two lipped. The ovary is superior and usually has two cells, and a single terminal style.

KEY

1. Corolla tube ending in spur or short basal pouch
  2. Corolla tube with short, swollen pouch at base
    3. Pnnl; fls pale yellow veined with red .... *Antirrhinum majus* (introduced Ma)
    3. Annl; fls pink .... *Misopates orontium* (Ma, Mi, I)
  2. Corolla tube extended into conical or cylindrical spur
    4. Capsule opening by longitudinal splits
      5. Lvs palmately veined, stalked, usu. broader than long *Cymbalaria*
      5. Lvs pinnately veined, sessile, not usu. broader than long .... *Linaria*
    4. Capsule opening by apical pores
      6. Stems erect .... *Chaenorrhinum*
      6. Stems procumbent, climbing or straggling .... *Kickxia*
1. Corolla tube without spur or pouch at base
  7. Corolla white or blue; fertile stamens 2.......................... *Veronica*
  7. Corolla yellow, pink, purplish or reddish-brown; fertile stamens 4
    8. Corolla dark reddish-brown (at least upper lip).......... *Scrophularia*
    8. Corolla not dark reddish-brown
      9. Corolla pink, tube inflated, spotted inside .... *Digitalis dubia* (Ma, Mi, endemic)
      9. Corolla not as above
        10. Corolla strongly two lipped
          11. Corolla with pink upper lip and white or yellow lower lip .... *Bellardia trixago* (Ma, Mi, I)
          11. Corolla either all yellow or all purplish-pink......... *Parentucellia*
        10. Corolla not strongly two lipped
          12. Corolla yellow
            13. Corolla 4-7mm........ *Sibthorpia africana* (Ma, Mi, I, endemic)
            13. Corolla 12mm or more.................................. *Verbascum*
          12. Corolla pink .... *Erinus alpinus* (Ma)

Antirrhinum majus Snapdragon; Llops, Conejitos; Löwenmaul; Muflier.
Misopates orontium Weasel's snout; Boca de dragó, Botons de gat; Acker-Löwenmaul; Tête de mort.
Digitalis Foxglove; Didalera, Boca de llop; Fingerhut; Digitale.

VERBASCUM Aaron's rod, Mullein; Candeleria; Königskerze; Molène.

1. Stamens 4.. *V. creticum* (Ma, Mi, I)
1. Stamens 5
  2. Some stamens with white haired filaments............ *V. thapsus* (Ma)
  2. Filaments of stamens with violet hairs

3. Calyx not exceeding 4.5mm
4. Basal lvs long stalked, dark green above .... *V. nigrum* (Ma)
4. Basal lvs sessile or very short stalked, grey or yellow tomentose .... *V. sinuatum* (Ma, Mi, I)
3. Calyx 6-9mm.. *V. boerhavii* (Ma)

SCROPHULARIA
1. Stems winged on angles .... *S. auriculata* (Ma, Mi, I)
1. Stems not winged
2. Annl; lvs up to 10cm, irregularly serrate ...... *S. peregrina* (Ma, Mi, I)
2. Pnnl; lvs 1-8cm, entire to twice pinnatisect
3. Cymes 1-3 fld; pedicels persistent, becoming spiny .... *S. ramosissima* (Ma, Mi, I)
3. Cymes 3-25 fld; pedicels not spiny................ *S. canina* (Ma, Mi, I)

Intermediates between the last two species occur here.

CHAENORRHINUM
Annl; lowest lvs red beneath.. .... *C. rubrifolium* ssp *formenterae* (Ma, I, endemic ssp)
Pnnl; all lvs green on both sides ....
C. origanifolium ssp *origanifolium* (Ma), ssp *crassifolium* (I, endemic ssp)

LINARIA Toadflax; Linaria; Leinkraut; Linaire.
1. Stems erect from a creeping rhizome; corolla white to pale lilac with violet veins .... *L. repens* (Ma)
1. Plant without creeping rhizome; corolla not as above
2. Corolla 12-30mm
3. Usu. pnnl; glandular-pubescent in infce .... *L. aeruginea* (Ma, ssp *pruinosa* endemic)
3. Annl; racemes glabrous
4. Calyx 4-5mm; corolla purplish-violet with white boss................... .... *L. pelisseriana* (Ma, Mi)
4. Calyx 7-12mm, fls white, pale yellow or violet, with orange boss
5. Fls subsessile; palate orange, spur violet... *L. triphylla* (Ma, Mi, I)
5. Pedicels 1-2mm, 4mm in fr.; palate and spur white .... *L. chalepensis* (Ma, Mi, I)
2. Corolla 4-9mm
6. Spur strongly curved.. *L. arvensis* (Ma)
6. Spur straight or slightly curved

7. Corolla pale yellow; spur 2-3.5mm .... *L. simplex* (Ma, I)
7. Corolla lilac-blue; spur up to 1mm.............. *L. micrantha* (Ma, I)

CYMBALARIA

1. Calyx 4-5mm; corolla 15-18mm .... *C. hepaticifolia* (Ma, ssp *balearica* endemic)
1. Calyx 2-2.5mm; corolla 8-15mm
2. Lvs 5-9-lobed.......................... *C. muralis* (Ma, Mi, I, introduced)
2. Lvs entire to 3-(occasionally 5-) lobed .... *C. aequitriloba* (ssp *aequitriloba* with violet corolla Ma, Mi: ssp *fragilis* with pale blue to whitish corolla endemic Mi)

KICKXIA

1. All lvs rounded at base or cordate (notch downwards)
2. Corolla yellow with deep purple upper lip .... *K. spuria* (Ma, Mi, I)
2. Corolla whitish, with violet upper lip and blue spotted boss............. .... *K. lanigera* (Ma,I)
1. At least some lvs arrow- or spear-shaped
3. Corolla 4-6mm; capsule 1.5-2mm .... *K. cirrhosa* (Ma, Mi)
3. Corolla 7-15mm; capsule 2.5mm or more
4. Spur strongly curved.......................... *K. commutata* (Ma, Mi, I)
4. Spur straight or nearly so .... *K. elatine* (I)

VERONICA Speedwell; -; Ehrenpreis; Véronique.

1. Fls in opposite axillary racemes
2. Annl; corolla 3-5mm, lilac or whitish .... *V. anagalloides* (Ma, Mi, I)
2. Pnnl (rarely annl); corolla 5-10mm, blue with violet veins............... .... *V. anagallis-aquatica* (Ma, Mi, I)
1. Fls solitary, axillary, or in terminal racemes
3. Fls white... *V. cymbalaria* (Ma, Mi, I)
3. Fls blue or occasionally pinkish
4. Capsule hairless or with marginal cilia only; glandular hairs absent
5. Capsule with ciliate margin
6. Corolla 2-3mm; capsule 3 x 3mm .... *V. arvensis* (Ma, Mi)
6. Corolla 8-12mm; capsule 6-7mm................... *V. persica* (Ma, I)
5. Capsule hairless .... *V. hederlfolia* (Ma, Mi, I)
4. Capsule hairy, glandular hairs present
7. Plant erect; capsule with glandular hairs only
8. Style 2mm .... *V. praecox* (?Ma)
8. Style 0.5mm.. *V. verna* (Ma)

7. Plant procumbent or decumbent; capsule with glandular and curled eglandular hairs .... *V. polita* (Ma, Mi, I)

PARENTUCELLIA

Fls yellow .... *P. viscosa* (Ma, Mi, I)
Fls pinkish-purple .... *P. latifolia* (Ma, Mi)

GLOBULARIACEAE

This is a small family with only one genus in Europe. The flowers could be taken for *Scabiosa* (Scabious; Escabiosa; Skabiosa; Scabieuse), which they resemble in the dense rounded terminal heads surrounded by an involucre of bracts, and in the pale mauve-blue colour.

The small flowers which make up the head differ from Dipsacaceae in the absence of a second, outer calyx-like structure, the epicalyx, and in the structure of the corolla, which is strongly two-lipped, the upper lip very short, and the lower long and 3-lobed.

GLOBULARIA Locally: Cossiada, Cebollada, Coronilla de fraile.

Twiggy shrub; heads 1-2.5cm .... *G. alypum* (Ma, Mi, I)
Herbaceous pnnl, often woody below, heads about 3.5cm .... *G. cambessedesii* (Ma, endemic)

ACANTHACEAE

Acanthus mollis (Bear's breech; Branca ursina, Hierba gigantia; Bärenklau; Acanthe) is cultivated here, and sometimes found apparently wild. (Ma, Mi)

OROBANCHACEAE

Only one genus occurs here. The inflorescence is a spike or raceme, occasionally branched, with flowers very like those of Scrophulariaceae but without chlorophyll or anything that looks like a leaf. The plants are parasitic, and leaves are replaced by yellowish or brownish scales more or less like the bracts in the inflorescence.

The host plant varies with the species, but it is often impossible to tell which is the host without digging the plant up, so the hosts are not given here.

In the key, calyx lobes, bracts and bracteoles must be distinguished. The calyx may be bell-shaped with four teeth, but is usually divided into two lateral segments, entire or bifid. Outside the calyx, and more or less fused to it, there may be a pair of bracteoles, and whether there is or not is an important distinguishing feature. Below the calyx and bracteoles (where present), there is a bract arising from the main stem.

The colours are important; they deteriorate rapidly after picking, and usually disappear completely on drying. Photographs are very useful here.

OROBANCHE Broomrape; Espárrago de loba, Frare, Hierba tora; Sommerwurz; Orobanche.

1. Each fl. subtended by bracteoles as well as a bract
   2. Anthers hairless ,or sparsely hairy at base; stigma white, cream, or pale blue.. *O. ramosa* (see below)
   2. Anthers hairy, stigma white or yellowish........... *O. lavandulacea* (I)
1. Fls without bracteoles
   3. Corolla shining dark red inside
      4. Corolla yellow outside, usu. with red veins; lower lip ciliate........... .... *O. gracilis* (Ma, I)
      4. Corolla usu. dark red or purple outside, lower lip not ciliate
         5. Calyx 7-10mm; corolla yellow at base............. *O. sanguinea* (Mi)
         5. Calyx 10-20mm; corolla not yellow .... *O. foetida* (Ma, Mi, I)
   3. Corolla not shining dark red inside
      6. Filaments inserted 7-15mm above base of corolla .. *O. latisquama* (I)
      6. Filaments inserted less than 6mm above base of corolla
         7. Corolla 20-30mm, broadening at mouth to open bell with large frilled and rounded lobes, usu. purple veined on a white or pale pink background .... *O. crenata* (Ma, Mi, I)
         7. Corolla 14-22mm, not much broader at mouth, lobes narrow, frilled or not, usu. white to pale yellow or cream tinged with violet
            8. Bracts 7-15mm, stamens inserted 2-3mm above base of corolla...... .... *O. minor* (Ma)
            8. Bracts 10-22mm, stamens inserted 3-5mm above base of corolla
               9. Stigma purple .... *O. loricata* (Ma, Mi, I)
               9. Stigma yellow to orange .... *O. hederae* (Ma, Mi, I)

O. ramosa has three ssp here: ssp *mutelii* (I) has corolla 18-22mm: ssp *nana* (Mi) and *ramosa* (Ma, Mi, I) have corollas 10-15mm, ssp *mutelii* usu. branched, with numerous fls, ssp *nana* (Mi, I) usu. simple with few fls. *O.*

crenata, parasitizes Leguminosae, and destroys *Vicia faba* (Broad bean; Haba; Saubohne; Fève) and other crops.

PLANTAGINACEAE

Plantago (Plantain; Plantatge, Llanten; Wegerich; Plantain), is the only genus here. Many species are well-known weeds of field and garden, giving them a familiarity reflected in a large number of regional names. But the holiday-maker can afford to set aside any prejudices, and look closely at the elegant spikes of rather small tubular four-petalled flowers, usually set in an orderly rosette of leaves. The calyx is four lobed, and there are four stamens; each flower is subtended by a bract.

PLANTAGO

1. Lvs opposite; infce axillary .... *P. afra* (Ma, Mi, I)
1. Lvs in basal rosette; infce a terminal spike
  2. Corolla tube hairy
    3. Leaves pinnatifid or bipinnatifid
      4. Annl, bnnl or pnnl; stock not or hardly branched; corolla lobes 1mm .... *P. coronopus* (Ma, Mi, I)
      4. Pnnl; stock usu. branched; corolla lobes 1.5mm or more .... *P. macrorhiza* (Mi)
    3. Lvs linear, entire or sparsely toothed
      5. Lvs fleshy .... *P. crassifolia* (Ma, Mi, I)
      5. Lvs not fleshy
        6. Annl, bnnl or pnnl; stock not or hardly branched; corolla lobes 1mm .... *P. coronopus* (Ma, Mi, I)
        6. Pnnl; stock usu. branched; corolla lobes 1.5mm or more .... *P. macrorhiza* (Mi)
  2. Corolla tube hairless
    7. Anterior sepals connate for more than half length
      8. Bracts and sepals densely long-hairy .... *P. lagopus* (Ma, I)
      8. Bracts and sepals hairless or very shortly hairy
        9. Lvs pubescent; flowering stem with up to 5 grooves .... *P. lanceolata* (Ma, Mi, I)
        9. Lvs densely covered with appressed hairs; flowering stem with more than 5 grooves .... *P. argentea* (?BI in FE)
    7. Anterior sepals free for more than half length
      10. Lvs egg-shaped or elliptical, hairless or pubescent, abruptly narrowed into stalk.................................... *P. major* (Ma, Mi, I)

PLATE 5. Zygomorphic flowers.

	Key page
Papaveraceae	
1. *Fumaria capreolata*	53
2. *F. parviflora*	53
Leguminosae	
3. *Calicotome spinosa*	71
4. *Genista lucida*	71
5. *Spartium junceum*	68
6. *Astragalus boeticus*	71
7. *A. balearicus*	71
8. *Psoralea bituminosa*	71
9. *Vicia leucantha*	72
10. *V. sativa*	72
11. *V. lutea*	72
12. *Lathyrus cicera*	73
13. *Ononis minutissima*	74
14. *Trifolium stellatum*	77
15. *Dorycnium hirsutum*	78
16. *Tetragonolobus purpureus*	79
17. *Anthyllis cytisoides*	79
18. *A. vulneraria* ssp *praepropera*	79
19. *A. tetraphylla*	79
20. *Coronilla juncea*	80
21. *Hedysarum coronarium*	80
Polygalaceae	
22. *Polygala rupestris*	88
Violaceae	
23. *Viola arborescens*	93
24. *V. suavis*	93
Boraginaceae	
25. *Echium plantagineum*	117
Labiatae	
26. *Teucrium asiaticum*	123
27. *T. subspinosum*	122
28. *T. cossonii*	122
29. *T. polium* ssp *polium*	123
30. *Scutellaria balearica*	121
31. *Phlomis italica*	121
32. *Ballota nigra*	123
33. *Stachys ochymastrum*	123
34. *Calamintha sylvatica*	124
35. *Micromeria filiformis*	124
36. *Mentha pulegium*	124
37. *Rosmarinus officinalis*	122

	Key page
Solanaceae	
38. *Hyoscyamus albus*	126
Scrophulariaceae	
39. *Verbascum sinuatum*	129
40. *Scrophularia peregrina*	129
41. *Misopates orontium*	128
42. *Linaria triphylla*	129
43. *Cymbalaria muralis*	130
44. *C. aequitriloba* ssp *aequitriloba*	130
45. *Kickxia spuria*	130
46. *Digitalis dubia*	128
47. *Veronica persica*	130
48. *V. hederifolia*	130
49. *Parentucellia viscosa*	131
50. *Bellardia trixago*	128
Orobanchaceae	
51. *Orobanche ramosa*	132
52. *O. ramosa*	132
53. *O. hederae*	132
Campanulaceae	
54. *Laurentia gasparinii*	137
Iridaceae	
55. *Gladiolus illyricus*	165
Orchidaceae	
56. *Limodorum abortivum*	196
57. *Orchis longicornu*	199
58. *O. coriophora*	199
59. *O. tridentata*	199
60. *O. laxiflora*	198
61. *Aceras anthropophorum*	197
62. *Barlia robertiana*	198
63. *Anacamptis pyramidalis*	197
64. *Serapias lingua*	199
65. *Ophrys speculum*	200
66. *O. lutea*	200
67. *O. fusca*	200
68. *O. bertolonii*	199
69. *O. tenthredinifera*	200
70. *O. apifera*	200
71. *O. bombyliflora*	200

3
16
18
23
22
8
24
25
14
15
20
5
4
17
7
6
21
12
19
11
30
10
9
37
33
13
31
35
29
36
26
32
34
46
27
28
2
53
1
52
49
38
41
40
50
39
48
45
47
51
61
55
65
64
42
54
43
63
67
71
68
69
44
66
59
70
56
62
57
60
58

10. Lvs linear to lance-shaped, densely hairy, gradually narrowed to stalk
 11. Fl. spikes long, slender, often interrupted .... *P. albicans* (Ma, Mi, I)
 11. Fl. spikes egg-shaped to short cylindrical, not interrupted.. *P. bellardii* (Ma, Mi, I)

P. coronopus Buck's horn plantain; Banya de bou, Cornicelis, Barballa, Estrella; Schlitz-Wegerich;-.
P. afra called here: Llavor de pua, Herba pucera, Botja, Afrontallaurador; in France Herbe aux puces.
P. lanceolata Ribwort; Cinc-nervis; Spitz-Wegerich; Herbe à cinq côtes.

CAPRIFOLIACEAE

These are shrubs or climbing plants with a tubular five-lobed corolla and 5 sepals and stamens. The ovary is inferior, and the fruit fleshy.

KEY
1. Woody climber; fls zygomorphic .... *Lonicera*
1. Erect shrub or herbaceous pnnl
 2. Lvs pinnate .... *Sambucus ebulus* (Ma, Mi, I)
 2. Lvs simple
  3. Fls in flat-topped, umbel-like clusters .... *Viburnum tinus* (Ma)
  3. Fls mostly in pairs .... *Lonicera*

Sambucus ebulus Danewort; Ebols, Saüc, Mielgo; Attich; Yèble, Hièble.
Viburnum tinus Laurestinus; Lloreret, Marfull, Tino; -; Laurier tin.

LONICERA Honeysuckle; Xuclamel, Madreselva; Geissblatt; Chèvrefeuille.
Climbing plant; fls strongly zygomorphic............ *L. implexa* (Ma, Mi, I)
Small shrub; fls nearly actinomorphic .... *L. pyrenaica* ssp *majoricensis* (Ma, endemic ssp)

VALERIANACEAE

The species occurring here are herbs with small or very small but numerous flowers, closely packed on a more or less regularly dichotomous inflorescence. The corolla is tubular, 5 or occasionally 4 lobed.

KEY

1. Fr. without pappus; stamens 2 or 3
  2. Fls actinomorphic, tiny, usu. pale blue .... *Valerianella*
  2. Fls zygomorphic, 8mm or more, pink with red markings.................. .... *Fedia cornucopiae* (Ma)
1. Fr. with a pappus; stamen 1 .... *Centranthus*

VALERIANELLA Lamb's lettuce, Cornsalad; Herba dels canonges, Valerianillas; Feldsalat; Doucette, Rampon.

1. Calyx in fr. very small (longest part much shorter than fr.)
  2. Calyx reduced to a very narrow rim, not or hardly toothed .... *V. microcarpa* (Ma, Mi, I)
  2. Calyx toothed; tooth over fertile cell much larger than others .... *V. dentata* (?Ma, ?I)
1. Calyx in fr. distinct (longest part nearly as long as fr. or longer)
  3. Calyx membranous, net-veined, with 5 up to 6 non-spiny teeth
    4. Bracts ovate and scarious .... *V. coronata* (?Ma)
    4. Bracts linear, green .... *V. eriocarpa* (Ma, Mi, I)
  3. Calyx densely hairy, usu. concealing veins, with 6-12 spiny teeth .... *V. discoidea* (Ma, Mi, I)

CENTRANTHUS

Fls red; corolla tube with pronounced spur .. *C. ruber* (Ma, I, introduced)
Fls pink; spur on corolla tube reduced to very small rounded projection .... *C. calcitrapae* (Ma, Mi)

C. ruber Dustman's breeches; Bàlsam de jardi, Amores mil, Centranto; Spornblume; Centranthe.

DIPSACACEAE

The flowers, or florets, in this family are very small, usually zygomorphic. They are clustered into a stalked head, each floret sessile on the flat or conical receptacle, which is surrounded by calyx-like involucral bracts. The receptacle may have hairs or scales between the florets. Each individual floret has its own calyx,and outside this a further epicalyx, known in this family as the *involucel*.

KEY

1. Plant robust (usu. more than 1m), spiny; opposite stem lvs connate at the base .... *Dipsacus fullonum* (Ma)

1. Plant usu. less than 1m, not spiny; stem lvs not connate
  2. Involucel with papery cup.. *Scabiosa*
  2. Involucel quadrangular, toothed, without papery cup
    3. Fls pink or lilac, outer fls distinctly larger than inner ....
    .... *Knautia integrifolia* (?Ma)
    3. Fls white or yellowish, outer not or hardly larger than inner............
    .... *Cephalaria squamiflora* ssp *balearica* (Ma, endemic ssp)

Dipsacus fullonum Teasel; Guitarra, Escorpidors, Peines; Karde; Cardère.

SCABIOSA

1. Pnnl; leaves mostly basal, entire .... *S. cretica* (Ma, Mi, I)
1. Annl; leaves not mostly basal, upper ones at least pinnate
  2. Involucel-tube with 8 longitudinal grooves ....
  .... *S. atropurpurea* (Ma, Mi, I)
  2. Involucel-tube smooth below, with 8 pits near the apex..................
  .... *S. stellata* (?Ma)

CAMPANULACEAE

The corolla in this family is tubular and five-lobed, generally actinomorphic (some botanists separate the zygomorphic species into a different family, Lobeliaceae). The calyx is (here) five-lobed, fused to the inferior ovary at the base, and there are five stamens. There is usually a milky latex; the leaves are generally simple and alternate.

KEY

1. Fls zygomorphic; anthers connate to form tube round style................
.... *Laurentia gasparinii* (Ma, Mi, I)
1. Fls actinomorphic; anthers free
  2. Corolla deeply 5-lobed, almost to base .... *Legousia*
  2. Corolla shallowly 5-lobed
    3. Fls numerous, in flat-topped cluster ....
    .... *Trachelium caeruleum* (Ma, introduced)
    3. Fls not clustered .... *Campanula*

LAURENTIA has been renamed and split since the publication of the FE. The Majorcan plant is '*Solenopsis minuta*' (L.) C. Presl., a perennial with lilac flowers. '*S. laurentia*' (L.) C. Presl. is an annual with bright blue flowers, found in Minorca and Ibiza.

LEGOUSIA

Lvs hairy; calyx lobes straight, about twice length corolla lobes .... *L. hybrida* (Ma, Mi)

Lvs hairless; calyx lobes curved, about 3 x length corolla... *L. falcata* (Ma)

CAMPANULA

Corolla 20mm; calyx with small lobes alternating with 5 larger lobes........ .... *C. dichotoma* (I)

Corolla 3-5mm; calyx with 5 lobes only .... *C. erinus* (Ma, Mi, I)

COMPOSITAE

As in Dipsacaceae, the flowers or florets are small and tubular, arranged on a receptacle surrounded by calyx-like involucral bracts. The calyx is much reduced, but sometimes enlarges in fruit to produce a pappus of hairs or scales, often adapted to facilitate seed dispersal. The receptacle may be smooth or honeycombed, and there may be scales, bracts or bristles between the flowers. The flowers have an inferior ovary and a single style surrounded by a tube formed by fused anthers.

The florets are of three main kinds: all three kinds may be male, female or hermaphrodite, but the sex is rarely important for the purposes of identification.

1) actinomorphic, with 5 short equal teeth
2) zygomorphic, without a ligule
3) zygomorphic ligulate, with the corolla tube drawn out at one side into a strap-like ligule, usually 5-toothed at the tip.

Familiar plants of this family in which the pattern may be seen are *Bellis perennis* (Daisy; Margarideta, Chirivita, Bellorita; Gänselblümchen; Pâquerette), which has yellow actinomorphic florets in the middle and white ligulate florets round the edge of the receptacle; and *Taraxacum officinale* (Dandelion; Dent de lléo, Bufallums, Amargón; Kuhblume, Löwenzahn; Pissenlit, Dent-de-lion), which only has yellow ligulate florets.

The family has a world-wide distribution. It is divided into two subfamilies, the Carduoideae with daisy type flowers (sometimes like a daisy with the 'petals' pulled off), and the Chichorioideae with dandelion-type flowers. It is essential to understand that the 'petals' of a daisy are in fact the ligules of tiny zygomorphic florets.

KEY

Plants without latex; central florets never ligulate; other florets ligulate or not .... Subfamily CARDUOIDEAE

Plants with latex; all florets ligulate........ Subfamily CICHORIOIDEAE

Subfamily CARDUOIDEAE

1. Ligulate florets absent, or very small and more or less concealed
 2. Florets yellow, greenish-yellow and white
 3. Heads 8mm or less in diam.
 4. All lvs entire or minutely toothed, usu. grey-tomentose
 5. Heads conspicuously bright yellow
 6. Dwarf shrubs, heads in dense corymb 1.5-8cm across...*Helichrysum*
 6. Herbs, heads separate
 7. Plant densely white woolly .... *Otanthus maritimus* (Ma, I)
 7. Plant not white woolly
 8. Hairy pnnl with sticky glands.............. *Jasonia glutinosa* (Ma)
 8. Glabrous green annual .... *Cotula coronopifolia* (Ma)
 5. Heads not conspicuously bright yellow (if florets bright yellow, then largely concealed by bracts)
 9. Dwarf shrub or woody-based pnnl .... *Phagnalon*
 9. Annl
 10. Involucral bracts aristate
 11. Rosette of greyish lvs with central cluster of almost stalkless greenish heads .... *Evax pygmaea* (Ma, Mi, I)
 11. Stems 5-35cm, usu. branched; rosette absent............ *Filago*
 10. Involucral bracts not aristate
 12. Fl. heads greenish; fr. heads covered in recurved hooks... *Xanthium*
 12. Fl. heads yellowish or brownish; fr. heads without hooks
 13. Head clusters terminal, without subtending lvs; involucral bracts not pouched... *Gnaphthalium luteo-album* (Ma, Mi, I)
 13. Head clusters subtended by lvs; at least some involucral bracts pouched at base
 14. Subtending lvs usu. longer than clusters; middle involucral bracts strongly saccate; bracts stellate in fr., not falling with achene.......................... *Logfia gallica* (Ma, Mi, I)
 14. Clusters not overtopped by subtending lvs; inner involucral bracts strongly saccate, each enclosing a peripheral floret and falling with the achene .... *Bombycilaena discolor* (I, ?Ma)
 4. At least some leaves obviously toothed or more deeply divided
 15. Whitish gey aromatic pnnl .... *Artemisia*
 15. Green annual, sometimes foetid, not aromatic

16. Prostrate plant, rooting at nodes .... *Gymnostyles stolonifera* (Ma, Mi)
16. Erect plant, not rooting at nodes
17. Heads in terminal clusters .... *Senecio*
17. Heads in terminal and axillary spike-like racemes .... *Ambrosia maritima* (Ma, Mi, I)
3. Heads 8mm or more
18. Plants without spines, except sometimes on involucral bracts
19. Aromatic bluish shrub with globular pale yellow heads .... *Santolina chamaecyparissus* (Ma, Mi, I)
19. Green herbs, sometimes foetid, not aromatic
20. Involucral bracts spiny .... *Centaurea*
20. Involucral bracts not spiny
21. Outer involucral bracts with spreading or recurved tips... *Inula*
21. Outer involucral bracts appressed to inner .... *Senecio*
18. Plants spiny, not only on involucral bracts
22. Lvs with linear lobes with small spiny tips............... *Centaurea*
22. Lf lobes not linear, plants thistle-like, very spiny
23. Inner involucral bracts spreading, ligule-like .... *Carlina*
23. Inner involucral bracts not spreading nor ligule-like
24. Heads bright clear yellow ...... *Carthamnus lanatus* (Ma, Mi, I)
24. Heads pale yellowish.................................. *Carduncellus*
2. Fls white, purple, pink or bluish
25. Plants without spines, except sometimes on involucral bracts
26. Involucral bracts conspicuous, shiny and papery
27. Stout pnnl; involucral bracts, inflated, shining and overlapping, producing a head suggesting a pine-cone.............................. .... *Leuzea conifera* (Ma, Mi, I)
27. Slender annl; inner involucral bracts, spreading, shining- pink, ligule-like .... *Xeranthemum inapertum* (Ma)
26. Involucral bracts not conspicuous, shiny and papery
28. Involucral bracts with terminal spine or hook, or pinnately or palmately fringed appendage, sometimes spiny
29. Involucral bracts hooked .... *Arctium tomentosum* (or possibly related sp) (Ma)
29. Involucral bracts spiny or fringed, not hooked
30. Pappus absent, or all achenes with pappus of 2 rows of hairs..... .... *Centaurea*
30. Pappus of several rows of long hairs on inner achenes, no pappus on outer achenes.......... *Cheirolophus intybaceus* (Ma, I)
28. Involucral bracts without spiny or fringed appendage

31. Robust pnnl; upper lvs usu. palmately lobed, heads pinkish-white, very narrow, 5 or 6 florets only, in dense, flat- topped cluster .... *Eupatorium cannabinum* (Ma)
31. Not as above; lvs not palmate, heads with numerous florets
32. Pnnl
33. Heads narrow cylindrical; involucral bracts silver-haired .... *Staehelina dubia* (Ma, Mi, I)
33. Heads ovoid, strongly constricted at apex .... *Mantisalca salmantica* (Ma)
32. Annl .... *Crupina*
25. Spiny plants; spines not confined to involucral bracts
34. Lvs whitish or pale grey tomentose
35. Heads solitary, globular, about 4cm diam.; lf spines green to white .... *Onopordum*
35. Heads less than 1cm diam., in clusters of 2-4; lf spines bright yellow .... *Picnomon acarna* (Ma)
34. Lvs not whitish or grey tomentose, sometimes slightly cottony
36. Outer involucral bracts longer than florets. incurved and deeply cut into spiny lobes (Chinese lantern effect) .... *Atractylis*
36. Outer involucral bracts not longer than florets
37. Lvs with white variegation on lamina adjacent to veins
38. Florets all more or less similar
39. Involucral bracts weakly spiny; heads 1.5cm .... *Tyrimnus leucographus* (Ma)
39. Involucral bracts with spreading stout spines and short lateral spines; heads 4-8cm .... *Silybum marianum* (Ma, Mi, I)
38. Outer florets enlarged and spreading .... *Galactites tomentosa* (Ma, Mi, I)
37. Lvs without white variegated lamina, veins sometimes white
40. Clustered heads encircled by spiny upper lvs, longer than heads .... *Notobasis syriaca* (Ma, Mi, I)
40. Heads not encircled by upper lvs (tip of lf may exceed head)
41. Heads distinctly bluish-purple or blue
42. Outer involucral bracts stout, leathery, not lf-like .... *Carduncellus*
42. Outer involucral bracts resembling upper lvs .... *Cynara*
41. Heads pinkish-purple to reddish
43. Heads globular to pear-shaped, 1cm or more .... *Cirsium*
43. Heads cylindrical, less than 1cm .... *Carduus*
1. Heads with ligular florets surrounding central tubular florets
44. Tubular florets purplish-pink, ray florets pale violet........... *Senecio*

44. Tubular florets yellow or orange, ray florets variously coloured
45. Ray florets violet to pale pinkish-white on front of ligule
46. Ray florets violet to violet-blue, longer than involucre......... *Aster*
46. Ray florets violet- to pinkish-white, usu. very small, shorter than involucre .... *Conyza*
45. Ray florets not violet to pale pinkish-white though sometimes tinged with these colours on back of ligule
47. Ray florets yellow or orange, or white, yellow or orange with a central band of yellow or black (not just deeper shade)
48. Heads 2-4mm diam. in flat-topped cluster; ray florets broader than long.................................... *Achillea ageratum* (Ma, I)
48. Heads more than 4mm, not usu. arranged as above; ray florets longer than broad
49. Ligules clearly bicoloured (with sharp junction between the two colours)
50. Ligules bright yellow at centre, white or pale yellow peripherally
51. Fr. ribbed, not winged .... *Leucanthemum paludosum* (I)
51. Fr. ribbed, with 2 or 3 narrow wings .... *Chrysanthemum*
50. Ligules black with a white or bluish spot centrally, orange peripherally .... *Gazania rigens* (Ma, introduced)
49. Ligules more or less uniformly coloured, at least on front
52. Involucral bracts in one row, excepting sometimes a few very short outer bracts
53. Fls yellow; a few short outer bracts commonly present.......... .... *Senecio*
53. Fls orange; short outer bracts absent .... *Calendula arvensis* (Ma, Mi, I)
52. Involucral bracts in more than 1 row
54. Fleshy pnnl (maritime)... *Inula*
54. Plant not fleshy
55. At least lower lvs coarsely toothed or lobed, or more deeply divided
56. Leaves glandular-hairy, especially on margin .... *Doronicum grandiflorum* (Ma)
56. Leaves without glandular hairs.............. *Chrysanthemum*
55. Lvs entire or shallowly toothed
57. Outer involucral bracts at least as long as ray-florets
58. Outer involucral bracts spine-tipped .... *Pallenis spinosa* (Ma, Mi, I)
58. Outer involucral bracts not spine-tipped .... *Asteriscus*

57. Outer involucral bracts shorter than ray-florets
59. Involucral bracts with long scarious tips.......... *Pulicaria*
59. Involucral bracts without long scarious tips
60. Fr. hairless; pappus hairs fused near base ...... *Dittrichia*
60. Fr. pubescent; pappus hairs not fused........................ .... *Solidago virgaurea* (Ma, I, ?native)

47. Ray florets white (sometimes tinged with violet or pink at tip or on back, or slightly with yellow at base)
61. Ray florets pure white
62. Fl. stem thickened below heads; outer achenes winged .... *Anacyclus clavatus* (Ma, I)
62. Fl stem not thickened below heads; achenes unwinged........... .... *Anthemis*
63. Corolla tube of disc florets spurred at base.......................... .... *Chamaemelum mixtum* (Ma, Mi)
63. Corolla tube of disc florets not spurred .... *Anthemis*
61. Ray florets tinged pink or violet at tip, or on back of ligule
64. Stem unbranched; lvs all more or less basal
65. Robust pnnl; heads 2-3.5cm .... *Bellis*
65. Slender annl; heads 8-12mm.. .... *Bellium bellidioides* (Ma, Mi, I)
64. Stems branched and leafy, at least in lower part
66. Much branched pnnl; ray florets in 2 rows.......................... .... *Erigeron karvinskianus* (Ma, introduced)
66. Slightly branched annl; ray florets in 1 row .... *Bellis*

Subfamily CICHORIOIDEAE (Those who despise 'yellow dandelions' can go straight to the end of the key for the coloured ones. But the very large range of flowers included in this group is remarkable, and well worth the trouble taken to identify them. Ripe fruit (*achenes*) are nearly always essential).

1. Flowers yellow
2. All lvs basal (flowering stems sometimes with a few small bracts)
3. Receptacle with scales.. *Hypochaeris*
3. Receptacle without scales
4. Some achenes with feathery hairs............................ *Leontodon*
4. Pappus of simple hairs or scales (or both)
5. Pappus of unequal yellowish hairs, some achenes with scales only .... *Hyoseris*
5. Pappus of white or brownish hairs on all achenes
6. Pappus hairs brownish, rigid .... *Hieracium*

6. Pappus white, soft
7. Involucral bracts and upper part stem with black glandular hairs .... *Aetheorrhiza bulbosa* (ssp *bulbosa* Ma, Mi, I: ssp *willkommii* and ssp *ramosissima* Ma, I, ssp *willkommii* endemic)
7. Involucral bracts and upper stem without black glandular hairs... .... *Taraxacum*
2. Flowering stem with at least one well-developed lf
8. Lvs very stiff and ferociously spiny .... *Scolymus*
8. Lvs spineless, or slightly prickly
9. Spiny 'hedgehog' plant ..... *Launaea cervicornis* (Ma, Mi, endemic)
9. Not a 'hedgehog' plant
10. Involucral bracts hardening and forming spreading star in fr.; pappus absent...................... *Rhagadiolus stellatus* (Ma, Mi, I)
10. Involucral bracts not forming star in fr.; pappus of hairs or scales
11. Achenes strongly compressed
12. Achenes beaked.. *Lactuca*
12. Achenes without a beak .... *Sonchus*
11. Achenes not compressed
13. Some achenes with pappus of scales.................................. .... *Hedypnois cretica* (Ma, Mi, I)
13. All achenes with a pappus of hairs
14. Receptacle with scales
15. Some pappus hairs feathery .... *Hypochaeris*
15. All hairs simple.. *Crepis*
14. Receptacle without scales
16. Some pappus hairs feathery
17. Involucral bracts in one row, united at base .. *Urospermum*
17. Involucral bracts in 2 rows or more
18. Plant hispid....................... *Picris echioides* (Ma, Mi, I)
18. Plant hairless or softly hairy .... *Scorzonera*
16. All pappus hairs simple
19. Shrubby plant; stems rush-like; leaves few .... *Chondrilla juncea* (Ma, Mi)
19. Plant not shrubby; stems not rush-like, usu. leafy
20. Outer involucral bracts narrow and thread-like, longer than florets .... *Tolpis barbata* (Ma, Mi)
20. Outer involucral bracts not as above
21. Involucral bracts with a white papery margin; achenes transversely grooved .... *Reichardia*

21. Involucral bracts without papery margin; achenes ribbed
22. Receptacle with pits surrounded by hairs..... *Andryala*
22. Recepacle without pits surrounded by hairs
23. Achenes abruptly truncated above .... *Hieracium*
23. At least inner achenes beaked .... *Crepis*
1. Flowers not yellow
24. Fls clear blue, not or hardly purplish
25. Heads solitary on almost leafless stem; involucral bracts papery...... .... *Catananche coerulea* (Ma)
25. Heads almost sessile in axils of upper lvs; involucral bracts green, hairy .... *Chichorium intybus* (Ma, Mi, I)
24. Fls pinkish, violet or purplish
26. Fls exceeding 2cm diam.; involucral bracts in 1 row .... *Tragopogon*
26. Fls very small; involucral bracts in several rows .... *Lactuca*

Some common names:
Ambrosia maritima American wormwood; Atárraga; Ambrosie; Ambrosie.
Achillea Yarrow; Alè de bou, Altarreina, Algerato; Scharfgarbe; Achilée.
Arctium Burdock; Aquipegotes, Bardana; Klette; Bardane.
Calendula Marigold; Llevamà, Clavellines de mort, Flamenquilla; Ringelblume; Souci.
Catananche caerulea -; Cigala, Hierba cupido; Rasselblume; Cupidone.
Chichorium intybus Chicory; Cama-roja, Amárgon, Almirón amargo, Achicoria; Zichorie, Wegwarte; Chicorée, Barbe de capucin.
Gnaphthalium Cudweed; Peludarda; Ruhrkraut; Immortelle des marais.
Galactites tomentosa: Card calapater, Card blanc locally.
Mantisalca salmantica: Cabeçuda, Balega, Alcabota and Algarabía locally.
Otanthus maritimus Cotton-weed; Algodonosa, Arañera in Castilian.
Santolina chamaecyparissus Lavender cotton; Camamilla de muntanya, Broida femella, Boja, Paciencia, Guardaroppa;-; Petit cyprès.
Silybum marianum Milk-thistle; Card blanc, Cártamo mariano; Mariendistel; Chardon Marie, Chardon argenté.

BELLIS Daisy; Pasqüeta, Chirivita; Gänselblümchen; Pâquerette.
Pnnl; lvs basal; (fl. late summer to autumn) .... *B. silvestris* (Ma, Mi)
Annl; usu. some lvs on lower stem; (spring flowering)........................... .... *B. annua* (Ma, Mi, I)

Bellis perennis, common in most of Europe, seems to be missing here.

ASTER

Annl or bnnl; ligules about same length as pappus .... *A. squamatus* (Ma, Mi, I, introduced)
Pnnl; ligules much longer than pappus .... *A. tripolium* (Ma, Mi, I)

CONYZA

Plant hairless or nearly so; ligules spreading .... *C. canadensis* (Ma, Mi, I, introduced)
Plant hairy, ligules held vertically .... *C. bonariensis* (Ma, Mi, I)

C. bonariensis here includes both the variant with white to reddish-brown pappus and pyramidal inflorescence '*C. bonariensis*' (L.) Cronq., and the plant with a yellowish pappus and cylindrical inflorescence '*C. floribunda*' Kunth.

FILAGO Cudweed; Herba cotonera, Padre e hijos; Fadenkraut; Cotonière.

1. Pappus absent, or of up to 5 caducous hairs.......... *F. congesta* (Ma, I)
1. Pappus present
 2. Leaves not apiculate; cluster of 20-40 heads not overtopped by subtending lvs .... *F. vulgaris* (Ma, Mi, I)
 2. Lvs apiculate; cluster of 8-20 heads overtopped by subtending lvs....... .... *F. pyramidata* (Ma, Mi, I)

HELICHRYSUM Everlasting flower; Flor de tot l'any, Amaranto amarillo, Perpetua amarillo; Strohblume; Immortelle.

1. Involucre cylindrical to bell-shaped just before flowering; bracts closely and regularly imbricate, outer leathery............ *H. italicum* (Ma)
1. Involucre ovoid to globular before flowering; bracts laxly imbricate, outer scarious
 2. Lvs usu. less than 25mm, sometimes aromatic .... *H. stoechas* (Ma, Mi, I)
 2. Lvs usu. more than 30mm, sometimes much more, not aromatic
  3. Lower lvs usu. about 7mm wide, paddle-shaped.......................... .... *H. ambiguum* (Ma, Mi, endemic)
  3. Lower lvs usu. 2-3mm wide, linear to lance-shaped...................... .... *H. rupestre* (Ma, I)

There are a lot of subspecies here, and indeterminate species; they are not easily sorted out.

PHAGNALON

1. Heads in clusters of 2-6............................ *P. sordidum* (Ma, Mi, ?I)
1. Heads solitary
  2. Outer involucral bracts spreading....................... *P. saxatile* (Ma, I)
  2. All involucral bracts appressed to involucre......... *P. rupestre* (Ma, I)

INULA

Lvs elliptical, not fleshy; ligulate florets absent .... *I. conyza* (Ma)
Lvs linear, fleshy; ligulate florets present........ *I. crithmoides* (Ma, Mi, I)

I. conyza Ploughman's spikenard; Coniza; Dürrwurz; Oeil-de-cheval.
I. crithmoides Golden samphire; Salsó, Fonoll marí, Hierba del cólico;-; Inule-faux.

DITTRICHIA

Stem lvs clasping stem; ray florets much longer than involucre, spreading .... *D. viscosa* (Ma, Mi, I)
Stem lvs not clasping stem; ray florets not longer than involucre .... *D. graveolens* (Ma, Mi, I)

D. viscosa is called locally Bufarrel, Mata mosquera, Altabaca.

PULICARIA Fleabane; Herba pucera, Herba de moix, Cunilago; Flohkraut; Pulicaire.

1. Pnnl; heads 15-30mm diam., ligules spreading
  2. Heads solitary or few.................................. *P. odora* (Ma, Mi, I)
  2. Heads in dense, more or less flat-topped cluster .... *P. dysenterica* (Ma, Mi, I)
1. Annl; heads about 10mm, ligules usu. erect
  3. Middle and upper lvs linear, rigid...................... *P. sicula* (Ma, Mi)
  3. Middle and upper lvs lance-shaped to elliptical, undulate .... *P. vulgaris* (Ma, Mi)

ASTERISCUS

Pnnl; ligules as long as involucral bracts .... *A. maritimus* (Ma, Mi, I)
Annl; ligules shorter than involucral bracts .... *A. aquaticus* (Ma, Mi, I)

XANTHIUM Cocklebur; Llufes, Abriülls, Cadillo; Betterläuse; Lampourde.

Annl; no spines except on fr. .... *X. strumarium* (Ma, Mi)
Pnnl; very spiny .... *X. spinosum* (Ma, Mi, I)

ANTHEMIS

1. Receptacle without scales peripherally, with scales centrally ....
.... *A. cotula* (Ma, Mi)
1. Receptacle with scales all over
   2. Ligules almost as broad as long .... *A. maritima* (Ma, Mi, I)
   2. Ligules much longer than broad .... *A. arvensis* (Ma, Mi, I)

A. cotula Stinking mayweed; Camamilla pudenta; Stinkende Hundskamille; Maroute.
A. arvensis Corn chamomile; Botons, Camamilla; Hundskamille; Camomille.

CHRYSANTHEMUM

Upper lvs entire or toothed, half clasping stem; ray florets bright yellow, like disc florets .... *C. segetum* (Ma, Mi)
All lvs finely divided, not clasping stem; ray florets as above or white with broad yellow band towards centre .... *C. coronarium* (Ma, Mi, I)

C. coronarium Crown daisy; Bolitx, Moixos, Pajitos, Corona de rey; Goldblume; Marguerite dorée.
C. segetum (Corn marigold; -; Saat-Wucherblume; Chrysanthème-des-blès) is said to be common here, but may have been confused with the all yellow form of *C. coronarium*. It does not seem to be common, if it occurs.

ARTEMISIA Wormwood; Donzell, Altamira, Abrótano; Beifuss; Absinthe.

Most leaves of flowering stems simple .... *A. caerulescens* (Ma)
Lvs on flowering stems all divided .... *A. arborescens* (Ma, Mi, I)

SENECIO Ragwort; -; Greiskraut; Séneon.

1. Tube and ray-florets both pinkish-violet ....
.... *S. leucanthemifolius* ssp *rodriguezii* (Ma, Mi, endemic ssp)
1. Tube and ray-florets (if present) bright yellow
   2. Plant densely white-tomentose .... *S. bicolor* (Ma, Mi, introduced)
   2. Plant not densely white-tomentose
      3. Head diam. 1cm or less
         4. Plant glandular-pubescent, sticky .... *S. viscosus* (?Mi)

4. Plant not glandular-pubescent, sticky
   5. Involucral bracts black-tipped................ *S. vulgaris* (Ma, Mi, I)
   5. Involucral bracts not black-tipped...................... *S. lividus* (Mi)
3. Head diam. more than 1cm
   6. Robust, woody pnnl; lvs linear, not fleshy .... *S. linifolius* (I)
   6. Plant not woody; lvs not linear, fleshy .... *S. leucanthemifolius* (Ma, I)

S. vulgaris Groundsel; Hierba cana, Suzón; Kreuzkraut; Séneçon.

CARLINA

Inner involucral bracts yellow to yellowish-brown .... *C. corymbosa* (Ma, Mi, I)
Inner involucral bracts purple .... *C. lanata* (Ma, Mi, I)

ATRACTYLIS

Annl; upper stem lvs not much longer than involucre .... *A. cancellata* (Ma, Mi, I)
Pnnl; upper stem lvs much longer than involucre........... *A. humilis* (?I)

CARDUUS (With *Cirsium*: Thistle; Card, Cardo; Distel; Chardon).

Stems with continuous prickly wing............... *C. tenuiflorus* (Ma, Mi, I)
Stems with interrupted prickly wing, unwinged just below heads .... *C. pycnocephalus* (Ma, Mi, I)

CIRSIUM

1. Stem unwinged
   2. Heads large, solitary .... *C. echinatum* (Ma)
   2. Heads small, in terminal corymbs of 2-4.............. *C. arvense* (Ma, I)
1. Stems winged .... *C. vulgare* (Ma, Mi, I)

ONOPORDUM Cotton thistle; Card del dimoni, Cardo borriquero; Eselsdistel; Onoporde.

Lvs with less than 8 pairs of lobes....................... *O. macranthum* (Ma)
Lvs with more than 8 pairs of lobes......................... *O. illyricum* (Ma)

CYNARA Cardoon; Cardo; Kardone; Cardon.

Cauline lvs 1-2 pinnatisect, margins revolute; achenes winged................ .... *C. humilis* (?I)
Cauline lvs 1-2 pinnatifid, margins flat; achenes unwinged.................... .... *C. cardunculus* (Ma, Mi, I)

CENTAUREA

1. Fls pinkish or purplish
 2. Appendages of middle involucral bracts with short, equal, palmately arranged spines
  3. Appendages with spines not more than 3mm... *C. aspera* (Ma, Mi, I)
  3. Appendages with spines 3-5mm.............................. *C. seridis* (I)
 2. Appendages of middle involucral bracts with long terminal spine, much shorter lateral spines .... *C. calcitrapa* (Ma, Mi, I)
1. Fls yellow
 4. Terminal segment summer lvs with 3 apical spines .... *C. balearica* (Ma, Mi)
 4. Lvs without spines at all times
  5. Spines of apical appendages of middle involucral bracts palmately arranged... *C. solstitialis* (BI in FE)
  5. Spines of apical appendages of middle bracts pinnately arranged
   6. Lateral spines few, at base of terminal spine
    7. Stem winged above; involucre 8-12mm .. *C. melitensis* (Ma, Mi, I)
    7. Stem not winged above: involucre 15mm.......... *C. hyalolepis* (I)
   6. Lateral spines numerous, forming spiny border almost to tip of terminal spine .... *C. collina* (I)

C. aspera Rough star-thistle in English; Bracera, Bovenaga locally.
C. calcitrapa Star-thistle; Abriülls, Cagatripa, Abrepuño, Cacítrapa; Stern-Flockenblume; Chausse-trappe.

CRUPINA

Heads with 3-5 florets .... *C. vulgaris* (Ma)
Heads with 9-15 florets.................................... *C. crupinastrum* (Ma)

CARDUNCELLUS

1. Stem lvs up to 6
 2. Outer involucral bracts spreading; pappus white............................ .... *C. monspelliensium* (I)
 2. Outer bracts appressed; pappus brownish .... *C. pinnatus* (Ma, I)
1. Stem lvs 10 or more
 3. Corolla yellowish... *C. dianius* (I)
 3. Corolla blue .... *C. caeruleus* (Ma, Mi, I)

SCOLYMUS

Spanish oyster plant; Tagarmina, Caderlina, Cardo zafranero, Cardo de olla; -;Epine-jaune.

2
8
41
34
31
28
13
5
12
30
39
40
27
14
32
7
15
22
3
35
36
1
24
23
14
20
19
18
10
38
16
33
9
25
21
4
11
17
26
29
6
37

PLATE 6. Spore-bearing plants and plants in which flower structure is not easily recognised.

	Key page
Sellaginellaceae	
1. *Selaginella denticulata*	23
Equisetaceae	
2. *Equisetum ramosissimum*	24
Adiantaceae	
3. *Adiantum capillus-veneris*	25
Aspleniaceae	
4. *Asplenium onopteris*	27
Polypodiaceae	
5. *Polypodium australe*	28
Polygonaceae	
6. *Rumex bucephalophorus*	37
Chenopodiaceae	
7. *Salicornia europaea*	40
Amaranthaceae	
8. *Amaranthus blitoides*	40
Euphorbiaceae	
9. *Euphorbia chamaesyce*	85
10. *E. dendroides*	87
11. *E. biumbellata*	87
Umbelliferae	
12. *Eryngium campestre*	103
13. *Bupleurum baldense*	104
Posidoniaceae	
14. *Posidonia oceanica*	157
Juncaceae	
15. *Juncus acutus*	167
Gramineae	
16. *Desmazeria marina*	177
17. *Cynosurus echinatus*	178

	Key page
18. *Lamarckia aurea*	178
19. *Briza maxima*	179
20. *Brachypodium phoenicoides*	180
21. *B. distachyon*	180
22. *Elymus elongatus*	181
23. *Aegilops ventricosa*	181
24. *A. geniculata*	181
25. *Lagurus ovatus*	184
26. *Polypogon monspeliensis*	185
27. *Phleum arenarium*	186
28. *Parapholis incurva*	186
29. *Stipa capensis*	187
30. *Cynodon dactylon*	188
31. *Paspalum paspalodes*	189
32. *Setaria verticillata*	190
33. *Pennisetum villosum*	190
34. *Hyparrhenia hirta*	190
Araceae	
35. *Arum pictum*	192
36. *Arisarum vulgare*	192
Lemnaceae	
37. *Lemna minor*	192
Typhaceae	
38. *Typha domingensis*	193
Cyperaceae	
39. *Scirpus holoschoenus*	194
40. *Schoenus nigricans*	194
41. *Carex flacca*	196

Stem with narrow, spiny interrupted wings .... *S. hispanicus* (Ma, Mi, I)
Stem with uninterrupted wings, 2mm wide at least.. *S. maculatus* (Ma, Mi)

HYOSERIS

Stem much thickened below head; peripheral florets usu. about 6 .... *H. scabra* (Ma, Mi, I)
Stem not thickened below heads; peripheral florets usu. 9-15 .... *H. radiata* (Ma, Mi, I)

UROSPERMUM

Heads pale yellow with blackish centre, back of outer ligules reddish; involucral bracts softly hairy .... *U. dalechampii* (Ma, Mi, I)
Heads uniform pale or bright yellow; involucral bracts bristly haired .... *U. picroides* (Ma, Mi, I)

HYPOCHAERIS Cat's ear;-; Ferkelkraut; Porcelle.

Involucral bracts densely hispid................ *H. achryophorus* (Ma, Mi, I)
Involucral bracts sparsely bristly on mid-vein .... *H. glabra* (Mi)

LEONTODON

Outermost ligules grey-violet beneath; pappus of scales .... *L. taraxacoides* (Ma, Mi, I)
Outermost ligules greenish beneath; pappus of hairs .... *L. tuberosus* (Ma, Mi, I)

SCORZONERA

Lvs simple, entire or toothed .... *S. hispanica* (Ma, I)
Lvs pinnate.. *S. laciniata* (Ma, Mi, I)

TRAGOPOGON

Fls pale violet; involucral bracts usu. slightly longer than outer florets....... .... *T. porrifolius* (Ma, I)
Fls pinkish; involucral bracts usu. much longer than outer florets .... *T. hybridus* (Ma, Mi)

T. porrifolius Salsify; Salsifí; Haferwurz; Salsifis.

REICHARDIA

Centre of head purplish-black .... *R. tingitana* (Ma, Mi, I)
Heads uniformly yellow................................ *R. picroides* (Ma, Mi, I)

SONCHUS Sow-thistle; Cerraja; Hasenkohl; Laiteron, Lierge.

1. Pnnl
   2. Heads in a loose corymb; involucre with dense yellow glandular hairs .... *S. arvensis* (Mi)
   2. Heads mostly solitary; involucre eglandular, sometimes cottony .... *S. maritimus* ssp *maritimus* (Ma, Mi, I)
1. Annl or bnnl
   3. Upper lvs with rounded auricles........................ *S. asper* (Ma, Mi, I)
   3. Lf auricles not rounded
      4. Lf lobes constricted to stalk-like base .... *S. tenerrimus* (Ma, Mi, I)
      4. Lf lobes not constricted at base or 0.......... *S. oleraceus* (Ma, Mi, I)

LACTUCA Lettuce; Lletuga, Lechuga; Lattich; Laitue.

1. Ligules purplish .... *L. tenerrima* (Ma)
1. Ligules yellow, sometimes slightly purple-tinged
   2. Achenes greyish .... *L. serriola* (Ma, Mi)
   2. Achenes black or blackish
      3. Stem lvs strongly decurrent, fr. and beak black ...... *L. viminea* (Ma)
      3. Stem lvs not decurrent, fr. black or blackish with white beak
         4. Margins of lvs entire; fr. half length beak .... *L. saligna* (Mi)
         4. Margins of lvs prickly; fr. about equalling beak ..... *L. virosa* (Ma)

TARAXACUM Dandelion; Dent de lléo, Amargón; Kuhblume; Pissenlit.
This is a difficult genus. As well as species produced by normal sexual fusion, there are species arising from seeds not produced in the ordinary way. Hundreds of species and subspecies have been described, including at least seven in these islands.

An easily recognized, constant species is: *T. obovatum* (Ma, I), with hammer-like processes on the tip of the outer bracts, and cottony stems.

CREPIS Hawk's beard; -; Pippau; Crépide.

1. Plant without stems; heads sessile in central rosette ..... *C. pusilla* (Ma)
1. Plant with stems
   2. Hds drooping in bud....................................... *C. foetida* (Ma, I)
   2. Hds erect in bud
      3. Stems numerous; lvs simple, all or nearly all basal .... *C. triasii* (Ma, Mi, endemic)
      3. Stems few; lvs entire to pinnate, cauline lvs several
         4. Stems branched mainly in lower half; outer involucral bracts green, very narrowly linear.................. *C. bellidifolia* (BI in FE)
         4. Stems branched mainly in upper half; outer involucral bracts with scarious margins, not very narrowly linear ...... *C. vesicaria* (Ma, I)

ANDRYALA

Heads numerous; involucre with spreading yellow hairs........................ *A. integrifolia* (Ma)

Heads few; involucre with felted white hairs.................. *A. ragusina* (I)

HIERACIUM

The problem here is the same as in *Taraxacum*.

H. elisaenum (Ma, endemic), with feathery hairs on leaf margins and under the midrib may be found in Majorca among many others more difficult to name.

Keys to genera of Monocotyledons (Angiospermae)

Theophrastus in 370 BC may have been the first botanist to draw attention to the differences between Monocotyledons and Dicotyledons. The technical distinction is the presence of one or two 'seed-leaves', or cotyledons, the first leaves of a plant already present in the seed before the development of a shoot.

By the time you want to identify a plant the seed-leaves will have usually disappeared. For practical purposes the distinction between the two great classes rests on other features – the basic six-petalled flower, with parallel veined, more or less linear leaves being characteristic of the monocotyledons, and the four or five-petalled flower with net-veined leaves of very variable shape being characteristic of the dicotyledons. Most gardeners without any notion of botanical classification, would separate tulips, daffodils, lilies, irises and grasses from delphiniums, daisies, poppies and primroses, just as Theophrastus did.

ALISMATACEAE

The Alismataceae are interesting as Monocotyledons very like some of the more primitive Dicotyledons, particularly Ranunculaceae. Because of this similarity it is thought that both may have arisen from a primitive ancestral group, the Ranales. The most important common feature is the head of free carpels, spirally arranged on a receptacle.

The flowers are made up of two whorls of three, the inner white or pinkish petals, and outer green sepals. The ovary is superior and stamens usually six. All the plants grow in water or wet places, and the leaves are all basal and entire.

KEY

1. Infce of simple whorls
2. Fls pale pinkish, up to 15mm diam... *Baldiella ranunculoides* (Ma, Mi)

2. Fls about 6mm diam., white .... *Damasonium alisma* (Ma)
1. Infce of compound whorls (branches of primary whorls themselves with whorls).. *Alisma*

ALISMA Water plantain; -;Froschlöfel; Plantain-d'eau.
Lvs ovate (fls open from 1-7pm) .... *A. plantago-aquatica* (Ma)
Lvs lanceolate (fls open 9am to 2pm)........... *A. lanceolatum* (Ma, Mi, I)

JUNCAGINACEAE

The only genus here is *Triglochin*, marsh herbs with basal linear leaves, and a dense many-flowered spike-like inflorescence. The flowers are small and greenish.

TRIGLOCHIN
1. Rhizomatous plant; all 6 carpels fertile .... *T. maritima* (?Ma)
1. Plant bulbous; only 3 carpels fertile.............................. *T. bulbosa*
2. Plant flowering March-May; pedicels more or less spreading in fr. .... *T. bulbosa* ssp *barrelieri* (Ma, Mi)
2. Plant flowering Sept.-Nov.; pedicels more or less appressed to rhachis in fr. .... *T. bulbosa* ssp *laxiflora* (Mi)

POTAMOGETONACEAE

POTAMOGETON is the only genus here. These are floating or submerged plants of fresh water (here), with small flowers in axillary spike. These have four sepal-like green segments and four stamens.

1. Floating lvs present
2. Submerged lvs all linear; floating lvs appearing jointed at junction with stalk, margins slightly decurrent down stalk ... *P. natans* (Ma, I)
2. Submerged lvs with at least a narrow translucent blade; floating lvs not jointed at junction with stalk, margins not decurrent
3. Floating lvs leathery; stipules 2-4cm .... *P. nodosus* (Ma, Mi)
3. Floating lvs thin, translucent; stipules 7-10cm............................ .... *P. coloratus* (Ma, Mi, I)
1. Floating lvs absent
4. Submerged lvs at least 2.5cm broad, usu. more, often wavy-edged
5. Lvs minutely toothed, with 11-13 longitudinal veins; stipules 3-8cm .... *P. lucens* (Ma)

5. Lvs serrated, with 3-5 longitudinal veins; stipules 1-2cm.. *P. crispus* (Ma, Mi)
4. Submerged lvs not more than 0.5cm broad, not wavy-edged
6. Stipules 6-17mm, tubular to above half way, pale brown .... *P. pusillus* (Ma, ?Mi)
6. Stipules 20-50mm, open and convolute with whitish margin.. *P. pectinatus* (Ma, Mi, I)

RUPPIACEAE

This family is rather like Potamogetonaceae, but the spikes of flowers are terminal, and there is no perianth. The carpels become long-stalked in fruit. The one genus is *Ruppia*, and the species here grow in salt-marshes.

RUPPIA

Peduncles spirally coiled in fr.............................. *R. cirrhosa* (Ma, Mi)
Peduncles wavy, not coiled, in fr...................... *R. maritima* (Ma, Mi, I)

POSIDONIACEAE

The only species here is *Posidonia oceanica*, locally called Altina or Alga marina. It lives and dies under the sea, and becomes noticeable chiefly when dead. The strap-shaped leaves with a rounded tip are washed up in enormous quantities on the beaches here to form rotting piles at the waters edge, or bleached and dried coils like wood-shavings among the vegetation of the higher parts of the beach. It is unusual to find a green part of the plant on the beach, but dead fragments of the rhizome may be found, with numerous brown bristly fibres, and pellets formed by felted masses of the same fibres, looking like horse-droppings washed by the sea.

Schnorkellers will see the living plant forming underwater meadows a little way offshore.

Local authorities here spend a lot of time, money and effort bulldozing tons of this plant from the sea shore to produce clean beaches for the tourists. It is quite useful all the same, and farmers carting off loads of *Posidonia* to fertilise their fields are a common sight.

ZOSTERACEAE

This family reduces the flower almost to the basic essentials, the male flower consisting of a single anther only, or sometimes with a rudimentary bract-like perianth, and the female flower being a single carpel. Identification usually depends on other features because flowers are often difficult to find.

The plants spread mainly vegetatively, and are grass-like with linear leaves sheathing at the bases, and shining brownish spreading rhizomes.

ZOSTERA Eel-grass; Pelota marina; Seegras; Varech, Goémon.
Lf sheaths entire, splitting only when old............. *Z. marina* (Ma, Mi, I)
Lf sheaths split .... *Z. noltii* (Mi, I)

ZANNICHELLIACEAE

This is yet another family of submerged aquatic plants. The flowers are minute and unstalked in the axils of the linear leaves.

KEY

1. Lvs 2-4mm wide, prickly toothed towards the apex (marine)............. .... *Cymodocea nodosa* (Ma, Mi, I)
1. Lvs up to 2mm wide, entire at apex (brackish water)
   2. Lvs all alternate .... *Althenia filiformis* (Ma)
   2. Lvs of sterile branches alternate, of fertile branches more or less opposite... *Zannichellia palustris* (Ma, Mi)

ZANNICHELLIA Horned pondweed; -; Teichfaden; Zannichellie.

NAJADACEAE

Also submerged, this one is at least easy to recognize. *Najas marina*, in spite of its name, grows in only slightly salty water and not the sea. The leaves are narrow and have broad spiny teeth. The flowers are small and axillary with no perianth or hardly any. (Ma)

LILIACEAE

This is a large family with rather variable habit but usually with very characteristic flowers with no green calyx and two whorls of three petals. The two whorls are often similar but not usually identical. The petals of the outer whorl often have a darker colour or stripe on the under surface. Some species have a tubular corolla or petals united at the base, with six lobes and no calyx. The leaves are usually linear with parallel veins.

KEY

1. Shrub or climbing pnnl, spiny
   2. Climbing plant, with hooked spines and tendrils; lvs leathery, heart-shaped or more or less triangular; fls in axillary racemes .... *Smilax aspera* (Ma, Mi, I)
   2. Plants not climbing, without tendrils, sometimes scrambling; lvs reduced to brownish scales; leathery leaf-like cladodes present (sometimes caducous, and not always present at fl. time)
      3. Upright shrub; cladodes ovate, spine tipped; fls and fleshy red globose fr. borne on upper surface cladode .... *Ruscus aculeatus* (Ma, Mi, I)
      3. Scrambling plant; cladodes linear, not always spiny; fls and black fleshy fruit borne in axils of brown scale-like lvs .... *Asparagus*
1. Plants not shrubby or climbing; spines absent
   4. Fls bright yellow, sometimes greenish outside .... *Gagea*
   4. Fls not bright yellow
      5. Corolla of free petals
         6. Fls umbellate, umbel enclosed in brownish papery spathe before flowering .... *Allium*
         6. Fls not umbellate; spathe absent
            7. Corolla pinkish-lilac, petals appearing as if united below, actually pressed closely together; fls solitary, almost stalkless .... *Merendera filifolia* (Ma, Mi, I)
            7. Corolla not pinkish-lilac, petals obviously free
               8. Fls white, 1cm or more, often with green stripe on petals .... *Ornithogalum*
               8. Fls bluish or purplish, less than 1cm .... *Scilla*
      5. Petals united at least at base
         9. Fls blue, or in racemes of which some fls are blue
            10. Corolla globular, with 6 very short lobes .... *Muscari*
            10. Corolla not globular, lobes equalling or exceeding tubular part .... *Aphyllanthes monspeliensis* (I)
         9. Fls not blue
            11. Fls yellowish, brownish or orange-red, in one-sided spike .... *Dipcadi serotinum* (I)
            11. Fls white or lilac-pink, not in a one-sided spike
               12. Fls white, often shaded or striped with other colours

13. Lvs present at time of flowering (spring); fls in branched infce .... *Asphodelus*
13. Lvs absent at time of flowering (autumn); fls in unbranched, spike-like infce .... *Urginea*

12. Fls lilac-pink
14. Lvs present at flowering time (spring); fls in more or less flat-topped cluster .... *Brimeura*
14. Lvs absent at time of flowering (autumn); fls solitary .... *Colchicum autumnale* (Ma)

Smilax aspera Salsaparilla; Aritja, Arinjol, Heura espinosa, Zarzaparilla; -; Liseron épineux.
Ruscus aculeatus Butcher's broom; Acebo pequeño, Arrayán morisco, Búfol, Alberuera; Besenheide; Buis pointu, Petit houx.
Urginea maritima Sea squill; Escilla, Ceborrancha; Meerzwiebel; Scille.

ASPHODELUS Asphodel; Albó, Cibolla, Asfódelo; Affodill; Alèd, Ciboule.
Lvs hollow, semicircular in section; fls pendant... *A. fistulosus* (Ma, Mi, I)
Lvs flat, keeled; fls not pendant......................... *A. aestivus* (Ma, Mi, I)

GAGEA
Stems, pedicels and petals glabrous; petals 1cm .... *G. nevadensis* (Ma)
Stems, pedicels and outer surface petals hairy; petals 15mm .... *G. arvensis* (Ma, I)

ORNITHOGALUM Star-of-Bethlehem; Lliri bord, Calabruixa, Lèche de pajarro; Milchstern; Ornithogale.

1. Raceme short, spreading; infce appearing umbellate, though not actually so; lvs with white stripe on upper surface
2. Subsidiary bulblets forming tufts of leaves around flowering bulb; capsule with 6 equally spaced angles .... *O. umbellatum* (Ma, I)
2. Subsidiary bulblets few or absent; capsule usu. with 3 distinct pairs of angles .... *O. collinum* (I)
1. Raceme elongated, not umbel-like; lvs without white stripe above
3. Ovary shining black, conspicuous................. *O. arabicum* (Ma, Mi)
3. Ovary green, inconspicuous................. *O. narbonense* (Ma, Mi, I)

SCILLA Squill; Jacinto; Blaustern; Scille.
Lvs 1-4mm wide, hairless .... *S. autumnalis* (Ma, Mi, I)
Lvs 5-25mm wide, hispid on margin............................ *S. obtusifolia* (I)

BRIMEURA

Racemes unilateral; fls 6-12, blue to violet .... *B. amethystina* (Ma)

Racemes more or less corymbose, with lowest pedicels longer than upper; fls 1-6(-8), lilac to pinkish-lilac or white .... *B. fastigiata* (Ma, Mi)

MUSCARI Grape hyacinth; Lliri d'ase, Candeler, Barralets, Capblau, Nazarenos; Traubel hyazinthe; Muscari.

1. Fertile fls blue; sterile fls, if present, similar or paler
 2. Autumn flowering; perianth pale blue .... *M. parviflorum* (Ma)
 2. Spring flowering; perianth blackish-blue .... *M. neglectum* (Ma, I)
1. Ferile fls yellow to brownish; sterile fls blue .... *M. comosum* (Ma, Mi, I)

ALLIUM Wild garlic, leek, onion, chives; Ajo, Porro bord, Porradell, Puerro, Cebolla, Cebollino; Knoblauch, Poree, Sommerzwiebel, Schnittlauch; Ail, Oignon, Poireau, Ciboulette.

1. Umbel with bulbils only; fls absent
 2. Lvs basal, not sheathing stem .... *A. nigrum* (Ma)
 2. Lvs sheathing lower half stem or more
  3. Spathe 1-valved, deciduous .... *A. vineale* (Ma, Mi, I)
  3. Spathe usu. 2-valved, persistent .... *A. sphaerocephalon* (Ma, I)
1. Umbel with fls; bulbils present or not
 4. Lvs basal or sheathing not more than 1/6 stem, parting from stem at about same level
  5. Umbel almost sessile in rosette of lvs .... *A. chamaemoly* (Ma, Mi, I)
  5. Umbel on obvious stem
   6. Lvs with cilia on at least part of margin
    7. Stamens about equalling petals .... *A. subvillosum* (Ma, ?Mi)
    7. Stamens up to 2/3 length petals............ *A. subhirsutum* (Ma, Mi)
   6. Lvs glabrous
    8. Leaves hollow, section semicircular .... *A. cepa* (Ma, Mi, I)
    8. Leaves flat or keeled, not tubular
     9. Stem not smoothly rounded throughout
      10. Stem sharply 3-angled .... *A. triquetrum* (Ma, Mi)
      10. Stem not sharply 3-angled
       11. Stem 2-edged above; fls lilac....... *A. senescens* (Is. Cabrera)
       11. Stem 2-edged below; fls greenish-white.. *A. victorialis* (Is. Cabrera)
     9. Stem smooth throughout
      12. Plant robust; lvs 15-60mm wide .... *A. nigrum* (Ma)
      12. Plant fairly slender; lvs 4-12mm wide ... *A. roseum* (Ma, Mi, I)

4. Lvs sheathing at least 1/4 stem, parting from it at different levels
  13. Spathe with 2 persistent valves
    14. Anthers yellow .... *A. pallens* (Ma)
    14. Anthers purple or reddish
      15. Anthers purple; stamens shorter than corolla tube ....
      .... *A. grosii* (I, endemic)
      15. Anthers reddish before opening; stamens longer than corolla tube .... *A. sphaerocephalon* (Ma, I)
  13. Spathe with 1 valve, persistent or not, or occasionally 2 deciduous valves
    16. Valve of spathe persistent; filaments all simple
      17. Stem smooth .... *A. cupani* (I)
      17. Stem angled, with 2 adjacent angles winged..........................
      .... *A. neapolitanum* (Ma, Mi)
    16. Valve(s) of spathe deciduous; filaments of 3 inner stamens tricuspid
      18. Lvs hollow .... *A. vineale* (Ma, Mi, I)
      18. Lvs more or less solid, sometimes with small hollow channels
        19. Corolla cup- or bell-shaped ...... *A. ampeloprasum* (Ma, Mi, I)
        19. Perianth egg-shaped
          20. Stamens not exserted .... *A. polyanthum* (Ma, Mi, I)
          20. Stamens exserted .... *A. commutatum* (Ma, Mi, I)

ASPARAGUS Asparagus; Esparaguera; Spargel; Asperge.

1. Cladodes solitary, stout, angled, spine-tipped, up to 7cm long............
.... *A. stipularis* (Ma, Mi, I)
1. Cladodes in clusters, slender, spine-tipped or not, not exceeding 1cm long
  2. Branches smooth; cladodes spiny-tipped; true lvs reduced to scales ....
  .... *A. acutifolius* (Ma, Mi, I)
  2. Branches ribbed; cladodes not spiny tipped; true lvs reduced to stout spine.. *A. albus* (Ma, Mi)

The young shoots of *Asparagus stipularis* are not spiny, and are gathered and eaten just like cultivated *Asparagus officinalis* ssp *officinalis*

AGAVACEAE

There are no native species, but *Agave americana* (Century plant; Aloe, Donarda, Atzavara, Cabuyá; Agave; Agave), a huge plant with massive rosettes of spiny glaucous leaves and solitary flowering stems up to 10m high is occasionally found growing apparently wild. Each stem has a branched terminal inflorescence of yellowish flowers. (Ma, I)

AMARYLLIDACEAE

The distinction between these plants and Liliaceae depends, like other matters of botanical classification, on how you define either. At present Amaryllidaceae are separated from Liliaceae on account of the inferior ovary, but the other constant feature is the presence of a papery spathe which encloses the immature flower or inflorescence. This occurs in *Allium* too, and this, and other features such as the umbellate inflorescence, have inclined some botanists to include *Allium* here in spite of the superior ovary.

KEY

1. Corona present
   2. Corona and corolla pure white; corona at least 20mm long with 12 teeth.. *Pancratium marinum* (Ma, Mi)
   2. Corona and corolla not both white; corona 5mm or less untoothed .... *Narcissus*
1. Corona absent
   3. Corolla bright yellow................................ *Sternbergia lutea* (Ma)
   3. Corolla white to creamish with green blotches.............................. .... *Leucojum aestivum* ssp *pulchellum* (Ma, Mi)

Pancratium marinum Sea daffodil; Azuncena d'arena, Amores mios; Narde; Lis de mer.
Leucojum aestivum Snowflake; Campanilla; Knotenblume; Nivéole.

NARCISSUS

1. Autumn flowering
   2. Lvs present at fl. time; fls in umbels of 3-7 .... *N. elegans* (Ma)
   2. Lvs absent at fl. time; fls usu. solitary......... *N. serotinus* (Ma, Mi, I)
1. Spring flowering .... *N. tazetta* (Ma, Mi, I)

DIOSCOREACEAE

The only species here is *Tamus communis* (Black bryony; Briònia negra,

Esparrago de caña; Schmerwurz; Herbe aux femmes battues). The stock is perennial, and the annual shoots arising from it have shining inverted heart-shaped leaves, with spike-like axillary clusters of very small green flowers, male and female on different plants, followed in autumn by bright red berries on the female plants. (Ma, Mi, I)

IRIDACEAE

This family, like Liliaceae and Amaryllidaceae, has 6 petals in two whorls of three. In Iridaceae these are nearly always fused at the base, and the flowers are often zygomorphic. There is usually a spathe, as in Amaryllidaceae and *Allium*. The most obvious feature that differentiates this family from the other two is that this has three stamens; they both have six. In addition, the style is very specialized. It is three-lobed and flattened, with the lobes often themselves deeply bilobed, or very large, winged and petal-like.

KEY

1. Style lobes much enlarged, with coloured, petal-like wings
  2. Plant with thick, fleshy horizontal rhizomes, often producing large numbers of stems over several square metres; height 60cm or more .... *Iris*
  2. Small bulbous plant with one or two stems; height up to 25cm .... *Gynandiris sisyrhinchium* (Ma, Mi, I)
1. Style lobes not enlarged, petal-like
  3. Fls zygomorphic
    4. Fls carmine-pink .... *Gladiolus*
    4. Fls pale yellow or greenish-yellow .... *Freesia refracta* (Ma, introduced)
  3. Fls actinomorphic
    5. Ovary below ground, perianth tube long and slender, stalk-like; stigma bright orange........ *Crocus cambessedesii* (Ma, Mi, endemic)
    5. Ovary above ground, perianth tube short; stigma yellow.... *Romulea*

IRIS Iris, Flag; Lliri, Lirio; Schwertlilie; Iris.

1. Fls yellow .... *I. pseudacorus* (Ma, Mi)
1. Fls white, bluish or violet
  2. Spathes entirely papery-white at time of flowering; fls pale blue........ .... *I. pallida* (Ma, Mi, I, introduced)
  2. Spathes green, with or without papery margin; fls white or violet

3. Spathe with brown-papery margin; fls violet or white tinged with violet.................................. *I. germanica* (Ma, Mi, I, introduced)
3. Spathe entirely green; fls pure white.... *I. albicans* (?introduced Ma)

The introduced *Iris* species often appear completely wild. The white form of *I. germanica* was formerly called '*I. florentina*'; according to FE it is often recorded in place of *I. albicans*, a species particularly associated with Muslim cemetaries. Apart from this, there is no record of *I. albicans* here, but it would be interesting to look out for it.

ROMULEA

Upper bract of spathe scarious, lower green with scarious margin .... *R. columnae* (Ma, Mi, I)
Upper bract with hyaline margin, lower entirely green .... *R. ramiflora* (Ma, Mi)

R. columnae – ssp *columnae* with almost glabrous filaments, lvs 3-8cm, occurs in two varieties, a commoner blue variety (Ma, Mi, I), and a distinctive small white var *lutea* (Ma only); ssp *rollii* with hairy filaments, lvs 15-30cm, occurs in Ma only.

GLADIOLUS Gladiolus; Espadelló, Lliri dels blats, Estoque; Siegwurz; Glaïel.

1. Anthers longer than filaments; seeds unwinged................................ .... *G. italicus* (Ma, Mi, I)
1. Anthers equalling filaments or shorter; seeds winged
2. Spike with 3-10fls, rarely branched .... *G. illyricus* (Ma, Mi, I)
2. Spike with 10-20fls, often branched .... *G. communis* (Ma)

There are two subspecies of *G. communis* here – both uncommon; ssp *communis* usually has pink or green veined basal sheaths and pink flowers, and is generally rather smaller than ssp *byzantinus* with dark red veins on the basal sheaths and red to purplish flowers. They are treated as separate species by some botanists.

JUNCACEAE

The only genus here is *Juncus* (Jonqueres, Jonques; Binse; Joncs). A lot of people who happily use keys to identify other plants give up when faced with rushes, sedges and grasses.

PLATE 7. Fruit and seeds.

	Key page
Pinaceae	
1. *Pinus pinea*	29
Cupressaceae	
2. *Juniperus phoenicea*	30
Taxaceae	
3. *Taxus baccata*	30
Ephedraceae	
4. *Ephedra fragilis*	31
Fagaceae	
5. *Quercus coccifera*	32
6. *Q. ilex*	32
Santalaceae	
7. *Osyris alba*	34
Polygonaceae	
8. *Rumex obtusifolius*	37
9. *R. conglomeratus*	37
10. *R. bucephalophorus*	37
Chenopodiaceae	
11. *Chenopodium album*	39
12. *Atriplex hastata*	39
Amaranthaceae	
13. *Amaranthus muricatus*	40
14. *A. albus*	41
Caryophyllaceae	
15. *Spergularia media*	45
Ranunculaceae	
16. *Clematis flammula*	49
17. *Ranunculus trilobus*	50
18. *R. muricatus*	50
19. *R. arvensis*	50
20. *R. sceleratus*	50
Papaveraceae	
21. *Papaver somniferum* ssp *setigerum*	52
22. *P. dubium*	52
23. *P. argemone*	52
24. *P. hybridum*	52
25. *Fumaria capreolata*	53
26. *F. parviflora*	53
Cruciferae	
27. *Clypeola jonthlaspi*	58
28. *Coronopus squamatus*	60
29. *Carrichtera annua*	57
30. *Succowia balearica*	57
31. *Rapistrum rugosum*	56
Resedaceae	
32. *Reseda luteola*	62
Platanaceae	
33. *Platanus orientalis*	64
Rosaceae	
34. *Rubus caesius*	65
35. *Sorbus domestica*	66

	Key page
Leguminosae	
36. *Ceratonia siliqua*	67
37. *Anagyris foetida*	69
38. *Lathyrus ochrus*	73
39. *Ononis crispa*	74
40. *Melilotus sulcata*	75
41. *M. messanensis*	75
42. *Medicago orbicularis*	75
43. *M. minima*	76
44. *Tetragonolobus purpureus*	79
45. *Hippocrepis balearica*	80
46. *Hedysarum spinosissimum*	80
Geraniaceae	
47. *Geranium dissectum*	81
48. *Erodium malacoides*	82
49. *E. ciconium*	82
Euphorbiaceae	
50. *Euphorbia pubescens*	86
51. *E. biumbellata*	86
Cneoraceae	
52. *Cneorum tricoccon*	88
Malvaceae	
53. *Malva sylvestris*	91
Ericaceae	
54. *Arbutus unedo*	105
Oleaceae	
55. *Phillyrea angustifolia*	110
Apocynaceae	
56. *Nerium oleander*	111
Asclepiadaceae	
57. *Gomphocarpus fruticosus*	111
58. *Vincetoxicum nigrum*	111
Rubiaceae	
59. *Galium tricornutum*	113
60. *G. verrucosum*	113
Scrophulariaceae	
61. *Veronica persica*	130
Compositae	
62. *Xanthium strumarium*	148
63. *Reichardia tingitana*	152
64. *Lactuca viminea*	153
Alismataceae	
65. *Damasonium alisma*	156
Liliaceae	
66. *Asparagus acutifolius*	162
67. *Ruscus aculeatus*	159
Iridaceae	
68. *Romulea ramiflora*	165
69. *Gladiolus illyricus*	165
Araceae	
70. *Arum pictum*	192

1
2
3
4
5
6
7
8
9
10
11
12
13
14
15
16
17
18
19
20
21
22
23
24
25
26
27
28
29
30
31
32
33
34
35
36
37
38
39
40
41
42
43
44
45
46
47
48
49
50
51
52
53
54
55
56
57
58
59
60
61
62
63
64
65
66
67
68
69
70

The three families have a lot in common superficially and tend to be lumped together. Rushes, however, are much easier than grasses. They are closely related to Liliaceae and in some ways very similar. The flowers are generally much smaller, with perianth segments greenish or brownish and rigid, not petal-like, but not really calyx-like either. They are called petals here for convenience. As in Liliaceae they are in two whorls. There may be three stamens or six.

The flowers are often in clusters, which are then arranged in panicles, with a bract under each panicle, cluster or flower. Sometimes the bracts have a green leaf-like point, occasionally long enough to look like a continuation of the main stem.

JUNCUS

1. Small annl; stem usu. less than 25cm
 2. Lf sheaths with auricles .... *J. pygmaeus* (Ma, Mi)
 2. Lf sheaths without auricles
  3. Lvs all basal; stems usu. 5cm or less, branched only from base .... *J. capitatus* (Ma, Mi)
  3. At least 1 stem lf present; stems usu. 8cm or more, much branched
   4. Inner 3 petals finely pointed, usu. longer than capsule .... *J. bufonius* (Ma)
   4. Inner 3 petals blunt to slightly pointed, not longer than capsule ...... *J. hybridus* (Ma, Mi, I)
1. Pnnl; stem of most species usu. more than 25cm
 5. Lvs with very sharp points
  6. Fls reddish-brown; capsule longer than petals... *J. acutus* (Ma, Mi, I)
  6. Fls straw-coloured; capsule not or hardly longer than petals............ *J. maritimus* (Ma, Mi, I)
 5. Lvs not sharply pointed
  7. Lvs with longitudinal septa (i.e. made up of more than 1 tube, easily seen in section of lf with lens)
   8. Stamens 6 .... *J. subnodulosus* (Ma, Mi, I)
   8. Stamens 3 .... *J. bulbosus* (?Ma)
  7. Lvs without longitudinal septa, (either solid or a single tube)
   9. Lvs with transverse septa
    10. Stems creeping with new shoots arising from nodes; capsule with beak 0.4-2mm................................... *J. fontanesii* (Ma, Mi)
    10. Stem erect to prostrate, but not rooting at nodes; capsule contracted to a point, but not beaked
     11. Lvs hardly compressed, usu. straight; transverse septa conspicuous .... *J. acutiflorus* (?Ma, ?Mi, ?I)

11. Lvs strongly compressed, usu. curved; transverse septa inconspicuous .... *J. articulatus* (Ma, Mi, I)

9. Lvs without transverse septa

12. Fl. stem without lvs; infce obviously terminal; bracts inconspicuous.. *J. subulatus* (Ma, Mi)

12. Fl. stem with 1-4 lvs; infce apparently lateral (ie overtopped by bract continuing line of stem).................. *J. inflexus* (Ma, Mi)

GRAMINEAE

Grasses are the most important of all plants. A book by John Christopher, originally published as *The Death of Grass* and later filmed as 'No Blade of Grass', gives a chilling and terribly plausible picture of a world in which grass is slowly exterminated by a virus.

Most people depend on staple foods derived from this family, whether rice, bread, porridge or pasta. Numerous other products such as maize, oats, barley, and sugar-cane add to the diet of man and beast, either harvested and processed, or as fresh grazing.

Many grass products are used for purposes other than food too, including thatching, basket-making, weaving, paper-making and rope-making.

Landscape depends on the visual impact of the green of grass, as well as its soil-binding properties, which prevent continual reshaping and destruction of land by wind erosion.

The main problem with identifying grasses is that the flowers are small, and it requires practice to dissect out the various parts needed to determine the genus and species. It is not very difficult though, and well worth the effort.

The flower of a grass is usually bisexual, with an ovary bearing two (occasionally one or three) stigmas, three (rarely two or one) stamens, and up to three minute scales or *lodicules* representing the perianth. All this is very tiny and difficult to see, and fortunately often not important.

The rest is more important. Each flower is enclosed in a pair of bracts called the *palea* and *lemma*. The lemma is the bigger, and enfolds not only the flower, but the margins of the palea too.

All this adds up to a *floret*. One floret or several, alternately and imbricately arranged along a very short, fine axis, the *rhachilla*, subtended at the base by two further bracts, the *glumes*, add up to the next unit, the *spikelet* (spklt here).

Spikelets may be arranged in many different ways upon variously branched inflorescences; and this final arrangement is largely responsible for the general habit of the grass, by which it will be recognized once it is familiar. The finer detail has to be used in the key because, as with so many plants, the characteristic appearance of a grass is usually difficult or impossible to describe exactly.

However arranged, the inflorescence is usually terminal on a jointed stalk or *culm*, which is sheathed by the lower end of a leaf, or leaves. At the point where the sheath leaves the culm and broadens into a blade, there is on the upper surface a transverse short membrane, or row of hairs or cartilaginous thickening called the *ligule*. The shape, size, and structure of the ligule is often very important for identification.

The following key departs from the pattern in the rest of the book in not giving the distribution of species which key out here. This is because all genera are briefly described after the key, including those with only one species, and the distribution is given there. It is hoped that this arrangement will help confirm (or otherwise!) identifications made from the key.

The key also rather stretches the meaning of the word *raceme*. Strictly this means an inflorescence which is unbranched, with the stalked flowers arranged along it with the youngest at the apex. It is here used to include part of an inflorescence arranged in this way. *Spike* is similarly stretched to include partial inflorescences. Spike-like means narrow and elongated with the pedicels and peduncles (where present) concealed, and includes inflorescences which are strictly racemes or panicles.

KEY

1. Culm ending in a single broad sheath, 4-5cm long, containing 2 or 3 partially fused spklts with long silky hairs .... *Lygeum spartum* p.188
1. Not as above.
 2. Spklts in groups of 3-7, with a central hermaphrodite spklt surrounded by 2-6 sterile spklts
  3. Lemmas of at least fertile spklts awned; awns long and conspicuous, softening and concealing outline of infce
   4. Infce becoming golden yellow; fertile and sterile spklts falling together .... *Lamarckia aurea* p.178
   4. Infce green or purplish-green; sterile spklts not falling .... *Cynosurus* p.178
  3. Lemmas all unawned (sterile spklts sometimes very small); outline of infce not concealed .... *Phalaris* p.186
 2. Spikelets not as above
  5. Spklts in pairs or threes, one sessile and hermaphrodite, one or (less often) two stalked and male or sterile
   6. Spikelets unawned throughout infce

7. Infce of terminal and axillary spike-like racemes .... *Hemiarthria altissima* p.191
7. Infce of usu. 4-10 spike-like racemes arising from near apex of stem.. *Digitaria sanguinalis* p.189
6. At least some spikelets awned
8. Infce a solitary, terminal raceme; spklts in lower part of infce unawned, upper ones long awned (5-8cm), awns forming a twisted spire .... *Heteropogon contortus* p.191
8. Infce not a solitary terminal raceme
9. Infce robust, loose,much branched; fertile spklts plump, shining, whitish, sterile ones smaller, purplish.. *Sorghum halepense* p.190
9. Infce robust or not; terminal branches narrow racemes; fertile spklts not plump and shining
10. Infce with lf-like bracts; branches with terminal paired racemes.. *Hyparrhenia hirta* p.190
10. Infce not leafy; several slender racemes arising close together from top of rhachis .... *Dichanthium ischaemum* p.191
5. Spklts not as in 5 above
11. Infce long and narrow, a single terminal spike or raceme
12. Glumes longer than florets
13. Spklts with 3 or more florets .... *Lolium* p.176
13. Spklts with 1 or 2 florets
14. Lateral spklts with 1 glume .... *Hainardia cylindrica* p.186
14. Lateral spklts with 2 glumes (side by side) ..... *Parapholis* p.186
12. Glumes shorter than florets
15. Spklts with back of lemma towards rhachis
16. Some lemmas in each spklt with geniculate dorsal awn
17. Pnnl.. *Avenula bromoides* p.183
17. Annl .... *Gaudinia fragilis* p.183
16. Lemma unawned, or awns not dorsal and geniculate
18. Lateral spikes with 1 glume
19. Annl; infce curved, up to 20cm....... *Psilurus incurvus* p.179
19. Pnnl; or annl exceeding 30cm, infce erect....... *Lolium* p.176
18. Lateral spklts with 2 glumes
20. Spklts more or less embedded in hollow in rhachis except at flowering .... *Narduroides salzmannii* p.177
20. Spklts not embedded in hollow in rhachis
21. Lower glume at least 3/4 upper
22. Pnnl .... *Brachypodium* p.180
22. Annl

23. Spklts 20-30mm .... *Brachypodium* p.180
23. Spklts less than 14mm
24. Stems up to 60cm; lemmas often awned .... *Micropyrum tenellum* p.177
24. Stems up to 30cm; lemmas unawned .. *Desmazeria* p.177
21. Lower glume less than 3/4 upper .... *Vulpia* p.176
15. Spklts with side of lemma towards rhachis
25. Spklts grouped 2 or 3 at each node of rhachis ... *Hordeum* p.181
25. Spklts solitary at each node of rhachis
26. Glumes with 2 prominent keels bearing tufts of hair .... *Dasypyrum villosum* p.181
26. Glumes keeled or not, if keeled, then keels without tufts of hair
27. Glumes broad, usu. tough, with 2 or more awns or teeth at apex
28. Glumes rounded on the back .... *Aegilops* p.181
28. Glumes with prominent keel(s) .... *Triticum* p.181
27. Glumes narrowed to a single point .... *Elymus* p.180
11. Infce not a single terminal spike or raceme
29. Infce of 2 or more spikes or racemes
30. Spikes or racemes arising several together from top of culm
31. Racemes usu. 2; spklts arranged 1 row .... *Paspalum* p.189
31. Spikes 3-6, with spklts in 2 rows .... *Cynodon dactylon* p.188
30. Spikes or racemes not all arising from top of culm
32. Rhachis flattened, with numerous very short appressed racemes .... *Stenotaphrum secundatum* p.189
32. Racemes branching from central rhachis
33. Raceme 2-sided; ligule of hairs...... *Aeluropus littoralis* p.188
33. Racemes 1-sided; ligule absent.............. *Echinochloa* p.189
29. Infce not composed of spikes or racemes, though often condensed and appearing like a spike or raceme (use lens to look for branches in infce)
34. Base of spklts surrounded by a ring of bristles
35. Bristles remaining after spklts have fallen; infce broad cylindrical, conspicuously soft, fluffy ... *Pennisetum villosum* p.190
35. Bristles falling with spklt; infce narrow cylindrical, not soft .... *Setaria* p.190
34. Not as above
36. Spklts of 1 hermaphrodite floret only
37. Glumes hairless or very minutely hairy

38. Robust maritime pnnl, often exceeding 1m; lvs inrolled, ligule 1-3cm long; infce dense, often more than 10cm........... .... *Ammophila arenaria* p.186
38. Annl or smaller pnnl, sometimes maritime; lvs various, ligule less than 1cm; infce rarely more than 10cm
 39. Annl; glumes hard, shiny, swollen at base; lemmas hairy with geniculate awn or hairless unawned, often both kinds in same spike.................. *Gastridium ventricosum* p.185
 39. Not as above
  40. Lemma awned
   41. Lemma hairless
    42. Infce lax, obviously branched; glumes free
     43. Spklts laterally compressed; glumes 1-veined, keeled .... *Agrostis* p.185
     43. Spklts dorsally compressed; glumes 3-5-veined, not keeled .... *Piptatherum* p.187
    42. Infce dense, elongated, spike-like; lower part glumes united .... *Alopecurus myosuroides* p.186
   41. Lemma hairy
    44. Awn from entire or minutely bilobed tip of lemma, more than 1.5, usu. more than 7cm .... *Stipa* p.187
    44. Awn 4mm from near base of lemma.......................... .... *Triplachne nitens* p.185
  40. Lemma not awned
   45. Ligule a row of hairs .... *Sporobolus pungens* p.188
   45. Ligule membranous
    46. Ligule truncated; spike falling entire... *Polypogon* p.185
    46. Ligule bluntly pointed, often torn; glumes persistent after spike has fallen......................... *Agrostis* p.185
37. Glumes with spines or hairs
 47. Glumes with hooked spines on veins .... *Tragus racemosus* p.189
 47. Glumes hairy
  48. Infce softly and conspicuously hairy
   49. Panicle egg-shaped; glumes narrow linear, unawned .... *Lagurus ovatus* p.184
   49. Panicle various, usu. lobed; glumes not linear, notched at tip, sometimes awned .... *Polypogon* p.185
  48. Infce not softly and conspicuously hairy
   50. Glumes with short, stiff, comb-like hairs along upper part keel .... *Phleum arenarium* p.186

50. Hairs not confined to keel .... *Lophochloa cristata* p.184
36. Spklt of 2 or more florets, sometimes including sterile florets
51. Ligule a row of hairs
52. Lemma or rhachilla with long silky hairs
53. Lemma with long hairs; rhachilla hairless ..... *Arundo* p.188
53. Rhachilla with long hairs; lemma hairless .... *Phragmites* p.188
52. Neither lemma nor rhachilla with long silky hairs
54. Spklts with 2 florets .... *Panicum* p.189
54. Spklts with many florets .... *Eragrostis* p.188
51. Ligule membranous
55. Lowest floret sterile
56. Infce not densely hairy; if cylindrical, up to 12cm (usu. less than 8cm); spklts with 2 sterile florets and single terminal hermaphrodite floret
57. Lemmas all unawned.............................. *Phalaris* p.186
57. Sterile lemmas awned.................. *Anthoxanthum* p.184
56. Infce densely silvery-hairy, cylindrical, more than 8cm long; spikelet of 2 florets, lower reduced to hyaline lemma, upper fertile
58. Long hairs arising below spklt; fertile lemma awnless, fringed at apex .... *Imperata cylindrica* p.190
58. Long hairs arising from rhachilla; fertile lemma with awn twice length lemma .... *Saccharum ravennae* p.190
55. Lowest floret male or hermaphrodite
59. Lemma broad, rounded, inverted heart-shaped.............. .... *Briza* p.179
59. Lemma usu. longer than wide, not inverted heart-shaped
60. Some lemmas awned; lemma at least 2mm beyond apex spklt
61. Annl
62. Glumes shorter than spklt; awn usu. terminal, straight
63. Lemma with 3 keels, at least in lower part............... .... *Vulpiella tenuis* p.177
63. Lemma rounded on back, or with 1 keel
64. Lemma awned from apex
65. Lower glume less than 3/4 upper, usu. much less .... *Vulpia* p.176
65. Glumes nearly equal ... *Micropyrum tenellum* p.177

64. Lemma awned from between 2 apical teeth............ .... *Bromus* p.179
62. Glumes almost as long as florets to longer
66. Spklts 10mm or more .... *Avena* p.183
66. Spklts up to 7.5mm
67. Lemma with 2 bristly points; awn dorsal, geniculate
68. Awn from above middle of lemma .... *Trisetum* p.184
68. Awn from below middle of lemma ...... *Aira* p.184
67. Lemma shortly bifid, awn from sinus or dorsal, not geniculate
69. Lemma with 3 veins, awn from sinus between apical teeth..................... *Avellinia michelii* p.185
69. Lemma with 5 veins, dorsal awn from near apex......................... *Lophochloa cristata* p.184
61. Pnnl
70. Lemmas awned from tip or near tip (sometimes some unawned, none awned from back)
71. Very robust, densely tufted plant, diam. usu. more than 1m; infce branched, reed-like; lemmas silky haired at base, short awned at tip .... *Ampelodesmos mauritanica* p.187
71. Small or fairly robust plants, less than 1m diam. infce various; lemmas not silky haired at base
72. Plant hairless or slightly hairy; lemmas equally awned .... *Festuca* p.176
72. Plant softly hairy all over; lower lemma unawned upper lemma shortly awned, awn becoming recurved when dry .... *Holcus lanatus* p.184
70. Lower lemma at least awned from back
73. Spklts usu. with 2 florets; lower lemma with awn from lower 1/3 back, upper awnless, or awned from back above middle .... *Arrhenatherum elatius* p.183
73. Spklts with up to 8 florets; all lemmas awned from about middle of back .... *Avenula* p.183
60. Lemma unawned, or with awn less than 2mm
74. Annl
75. Spklts with 2 florets .... *Aira* p.184
75. Some or all spklts with more than 2 florets

76. Panicle dense, ovoid or cylindrical.......................... .... *Lophochloa cristata* p.184
76. Panicle lax with obvious branches
77. Panicle branches rigid
78. Infce with many divaricate branches; lemma strongly keeled on all 3 veins............................ .... *Cutandia maritima* p.177
78. Infce sparingly branched, branches not divaricate; lemma smooth or slightly keeled above on back
79. Glumes not keeled ... *Micropyrum tenellum* p.177
79. Glumes strongly keeled.......... *Desmazeria* p.177
77. Panicle branches slender, flexuous
80. Pedicels thickened distally .... *Sphenopus divaricatus* p.177
80. Pedicels not thickened distally
81. Lemma awned............ *Avellinia michelii* p.185
81. Lemma unawned.......................... *Poa* p.177
74. Pnnl
82. Glumes almost as long as rest of spklt to longer
83. Lemmas all unawned; terminal floret sterile, club-shaped .... *Melica* p.179
83. Upper lemma with 2mm awn near tip, becoming recurved on drying; terminal floret usu. male, not club-shaped........................ *Holcus lanatus* p.184
82. Glumes much shorter than spklt, usu. about equalling lowest floret
84. Lemma with 3 veins .... *Koeleria macrantha* p.183
84. Lemma with 5 veins
85. Glumes and palea awned...... *Sesleria insularis* p.179
85. Glumes and palea unawned
86. Lemma rounded on back, or weakly keeled near apex
87. Midrib of lemma not reaching tip..................... .... *Puccinellia distans* p.178
87. Midrib of lemma reaching tip, sometimes excurrent... *Festuca* p.176
86. Lemma strongly keeled throughout length
88. Spklts in dense, one-sided masses at ends of branches .... *Dactylis glomerata* p.178
88. Spklts in loose or fairly dense panicle, not one-sided .... *Poa* p.177

The following descriptions are not intended to be complete, and do not necessarily apply to species not occurring in the Balearic Islands.

FESTUCA Fescue;-; Schwingel; Fétuque.
Pnnls; branched infce, glumes rounded on the back; spklts with 3 or more florets – usu. more. Lemmas often awned. Glumes equal or slightly unequal, shorter than lowest lemma.

Basal lvs at least very fine, thread-like, infolded............. *F. ovina* (?Ma)
Basal lvs 3-12mm wide, flat .... *F. arundinacea* (Ma, Mi)

LOLIUM Rye-grass; Ballico; Lolch; Ivraie.
Infce spike-like, with flattened spklts in 2 alternating rows, edgeways to rhachis.

1. Lemma becoming swollen, ovate when mature .... *L. temulentum* (Ma, Mi)
1. Lemma not swollen when mature, lance-shaped
   2. Pnnl; lvs flat or folded when young..................... *L. perenne* (Ma)
   2. Annl; lvs rolled in young shoots
      3. Stems with 4-5 nodes; spklt usu. with 11 or more florets................ .... *L. multiflorum* (Ma, Mi, I)
      3. Stems with 2-4 nodes; spklt with 11 florets or less .... *L. rigidum* (Ma, Mi, I)

VULPIA Fescue; -; Fuchsschwingel; Vulpie.
Annls; infce generally slender,unbranched or sparingly branched; only *V. geniculata* here with fairly robust much-branched infce. Lemmas all long-awned, glumes very unequal, lower not more than 3/4 length upper, generally much less.

1. Florets opening normally; anthers usu. 3, 2-5mm .... *V. geniculata* (Ma, Mi)
1. Florets cleistogamous, or anthers just protruding, 2mm or less
   2. Upper glume 10-14mm without awn, awn 4-6mm long; lower glume minute .... *V. fasciculata* (Ma, Mi, I)
   2. Lower glume at least 1/6 length of upper
      3. Lemmas long-ciliate .... *V. ciliata* (Ma, Mi, I)
      3. Lemmas not long-ciliate
         4. Lower glume 2.5-5mm, 1/2-3/4 length upper .... *V. bromoides* (Ma, Mi)
         4. Lower glume 0.5-3mm, less than 1/2 length upper
            5. Infce not fully exserted from uppermost shesth; upper glume unawned .... *V. myuros* (Ma)

5. Infce fully exserted from uppermost sheath; upper glume with awn usu. 2-3 x length glume........................ *V. muralis* (Ma, I)

MICROPYRUM

M. tenellum

Rather variable small annl, unbranched or sparingly branched. Spklts with 3-14 florets, with chaffy, awned or unawned lemma with hyaline margin. (?Ma)

NARDUROIDES

N. salzmannii

Small, rigid, rarely branched annl. Spklts are appressed into depressions in the rhachis. (Ma)

DESMAZERIA

Hairless annls, usu. small with rather rigid infce, spike-like above, sometimes branched below, branches hardly spreading.

Infce usu. unbranched; lower glume 2-3mm, acute; upper blunt .... *D. marina* (Ma, Mi, I)

Infce usu. branched below; lower glume 1.3-2mm; both glumes acuminate .... *D. rigida* (Ma, Mi, I)

CUTANDIA

C. maritima

Plant of maritime sands rather like *Desmazeria rigida*, only larger and more branched. Lemma strongly keeled on all 3 veins. (Ma, Mi, I)

SPHENOPUS

S. divaricatus

Smallish maritime annl, with much-branched panicle of small (2-3mm) spklts, pedicels filiform, gradually thickened distally. In spite of very fine structure, panicle is wiry and not floppy. (Ma, Mi, I)

VULPIELLA

V. tenuis

Maritime annl, with rigid branches, pedicels sharply angular. At maturity branches fold upwards to produce rather closed infce. (Ma, Mi)

POA Poa; Hierba depunta, Espiguilla; Rispengras; Paturin.

Annl or pnnl hairless grasses. Panicles more or less pyramidal in shape, generally loose. Spklts compressed, usu. with 2-5 florets, occasionally viviparous. Glumes keeled and awnless. Lemma usu. with tuft of cottony hairs at base and hyaline, awnless tip.

1. Spklts viviparous
2. Stem bulbous at base .... *P. bulbosa* (Ma, Mi, I)
2. Stem not bulbous at base .... *P. angustifolia* (Ma)
1. Spklts not viviparous
3. Annl; some lvs usu. transversely wrinkled; lower branches panicle deflexed after flowering .... *P. annua* (Ma, Mi, I)
3. Pnnl; lvs not transversely wrinkled; panicle branches not deflexed after flowering
4. Stem bulbous at base; glumes widest below middle .... *P. bulbosa* (Ma, Mi, I)
4. Stem not bulbous at base; glumes widest at or near middle
5. Ligule of uppermost lf not more than 3.5mm
6. Plant with long creeping rhizomes.............. *P. angustifolia* (Ma)
6. Plant not rhizomatous .... *P. flaccidula* (Ma)
5. Ligule of uppermost lf 5-10mm .... *P. trivialis* (Ma, Mi)

PUCCINELLIA

P. distans

Pnnl saltmarsh plant with more or less pyramidal, loose panicle, branches becoming deflexed after flowering. Spklt much as *Poa*. (Ma, Mi, I)

DACTYLIS

D. glomerata Cocksfoot; Knäuelgras; Dactyle pelotonné.

Rather coarse pnnl grass, spklts aggregated into dense, more or less lobed, clusters at ends of branches. (Ma, Mi, I)

CYNOSURUS Dog's tail; -; Kammgras; Cynosure.

Panicles dense, bristly, more or less ovate, one-sided. Outer spklts sterile, with comb-like spreading lemmas.

Lvs 3-9mm wide; lemmas of sterile spklts similar.... *C. echinatus* (Ma, Mi)

Lvs 1-3mm wide; upper lemmas much shorter and wider than lower .... *C. elegans* (Ma)

LAMARCKIA

L. aurea

Annl, dense one-sided infce with long awns, becoming bright golden yellow. Arrangement of spklts much as *Cynosurus*. (Ma, Mi, I)

PSILURUS

P. incurvus

Annl; spklts sessile, alternate, appressed to incurved, narrow axis. (Ma, I)

BRIZA Totter-grass, Quavers; Ballaruga, Belluguets, Pendientes, Tembladera; Zittergras; Amourette, Brize.

Annl. Spklts broad heart-shaped, pendulous, usu. purplish, on long slender pedicels.

Spklts 1-12, 10-25mm .... *B. maxima* (Ma, Mi, I)
Spklts numerous, 3-5mm .... *B. minor* (Ma, Mi, I)

SESLERIA

S. insularis

Small pnnl, stems slender. Infce spike-like, usu. dense. Glumes awned, lemma with broad 3-5-toothed tip. (Ma)

MELICA

Pnnl; spklts of 2-4 florets, rhachilla terminating in club-shaped mass of sterile lemmas. Glumes covering much of spike. Infce slender, with pendulous spklts, or more or less dense and spike-like.

1. Lemma hairy
 2. Panicle lax, usu. few-fld; lemma blunt, long-ciliate on proximal 2/3, hairless distally .... *M. bauhinii* (Ma)
 2. Infce spike-like, dense; lemma acute, long-ciliate to apex .... *M. ciliata* (Ma)
1. Lemma hairless .... *M. minuta* (Ma, Mi)

BROMUS

Annl or bnnl, usu. lax infce with long, many-fld spklts, drooping or erect on (usu.) long pedicels. Lemma usu. conspicuously long-awned.

1. Spklts parallel sided or V-shaped, wider at the apex; lower glume with 1 vein, upper with 3
 2. Lemma at least 20mm
  3. Panicle usu. lax, spreading; glumes with hyaline margins .... *B. diandrus* (Ma, Mi, I)
  3. Panicle usu. dense, stiffly erect; glumes mostly hyaline, except veins .... *B. rigidus* (Ma, Mi, I)
 2. Lemma up to 20mm
  4. Panicle drooping; most branches of panicle as long as spklts or longer .... *B. sterilis* (?Ma, ?Mi, ?I)
  4. Panicle erect; most branches of panicle shorter than spklts, or spklts sessile

5. Panicle lax, branches 10mm or more; lemma at least 3mm wide .... *B. madritensis* (Ma, Mi, I)
5. Panicle very dense; branches 1-10mm; lemma less than 3mm wide
  6. Lemma up to 1.5mm wide; 1-2 spklts on each branch of panicle..... .... *B. fasciculatus* (Ma, Mi, I)
  6. Lemma at least 2mm wide; 4-5 spklts on each branch of panicle..... .... *B. rubens* (Ma, Mi, I)

1. Spklts more or less elliptical, tapering towards apex; lower glume with 3-7 veins, upper with 5-9
  7. Lf sheaths with scattered hairs only
    8. Annl; lemma rounded dorsally, usu. awned .... *B. arvensis* (?Mi)
    8. Pnnl; lemma strongly keeled, awn short or 0 .... *B. wildenowii* (I)
  7. Lf sheaths hairy; awn of lemma usu. not slender and straight (and if so, then 3-7mm)
    9. Width of spklts 6mm or less
      10. Width of spklt 4-6mm; awn stout, straight, 4-11mm .... *B. hordeaceus* ssp *hordeaceus*
      10. Width of spklt 2-3mm; awn slender, straight, 3-7mm.................. .... *B. hordeaceus* ssp *thominii* (Ma, Mi, I)
    9. Width of spklt 6mm or more; awn twisted at base
      11. Lemma 11-20mm .... *B. lanceolatus* (Ma, Mi)
      11. Lemma 8-11mm................................ *B. squarrosus* (Ma, Mi)

BRACHYPODIUM False-brome; Fenàs, Fenazo; Zwenke; Brachypode.
Long spklts with many florets, inserted edgeways to rhachis.

1. Leaves hairy, not convolute
  2. Annl; lvs glaucous, twisted; spklts laterally compressed .... *B. distachyon* (Ma, Mi, I)
  2. Pnnl; lvs yellowish green, drooping; spklts not compressed .... *B. sylvaticum* (Ma, Mi)
1. Leaves hairless, often convolute
  3. Lvs up to 10cm; racemes of 1-7 crowded spklts, each up to 30 mm .... *B. retusum* (Ma, Mi, I)
  3. Lvs 10-40cm; racemes of 5-12 spklts, each up to 80 mm .... *B. phoenicoides* (Ma, Mi, I)

ELYMUS Lyme grass; -; Haargerste; Elyme.
Flattened spklts inserted broadside to rhachis in 2 alternating rows; inflce usu. rigid, elongated

1. Plant with long rhizomes, not forming dense tufts
  2. Rhachis fragile, breaking up at maturity .... *E. farctus* (Ma, Mi, I)

2. Rhachis tough, not breaking up at maturity
 3. Glumes acute, mucronate or shortly awned
  4. Lvs with crowded veins and rigid spinules on upper surface .... *E. pycnanthus* (Ma, Mi, I)
  4. Veins on upper surface lvs not crowded, spinules absent, sometimes hairy on veins .... *E. repens* (Ma, Mi, I)
 3. Glumes without awn or mucro, not very acute .... *E. pungens* (Ma, Mi)

1. Plant tufted, without rhizomes (lvs leathery with spinules) .... *E. elongatus* (Ma, Mi, I)

E. repens Couch grass; Grama del norte; Gemeine Quecke.

AEGILOPS

Infce a spike; spklts few, sessile, with large, broad, tough glumes, usually with several long straight awns.

1. Infce 1-2cm long; spklts 2-5, upper slender and sterile .... *A. geniculata* (Ma, Mi, I)
1. Infce 4-6cm long; spklts 4-10; spklts at base often vestigial
 2. Fertile spklts 4-7, scarcely swollen; vestigial spklts 2-3..................... *A. triuncalis* (Ma)
 2. Fertile spklts 5-10, usu. much swollen; vestigial spklts 0-2 .... *A. ventricosa* (Ma, Mi)

TRITICUM Wheat; Blat; Weisen; Blé.

These are cultivated species of wheat, really outside the range of this book. Spklt rather like that of *Aegilops*, but the broad, tough glumes are keeled and unawned, the lemmas often awned. The spikes are generally dense, erect and rigid.

DASYPYRUM

D. villosum

Hairless annl, with compressed two-sided spike, 4-12 x 0.6-2cm. Glumes truncated, with 2 leathery ciliate keels, lemma awned, awn up to 6cm. (Mi)

HORDEUM Barley; Espiga; Gerste; Orge.

Dense, long awned, spike-like infce, spklts in alternate groups of 3, broadside on to rhachis.

PLATE 8. Cultivated plants. Very many species of plant are cultivated here as crops and for ornament, and it would need another book to do justice to them all. Important crops not illustrated here include potatoes (a main export of Majorca, with three crops a year), strawberries, citrus fruits of all kinds, French beans (including haricots, which you may see being threshed in the traditional way), tomatoes, peppers and aubergines.

Crops which occasionally establish themselves in the wild will be found on the preceding pages.

		Key page
1.	*Ficus carica* (Moraceae) Fig; Figuera, Higuera; Feigenbaum; Figuier	33
2.	*F. elastica* Indiarubber fig	
3.	*Capparis spinosa* (Capparidaceae) Caper; Alcaparra; Kapernstrauch; Caprier	53
4.	*Aeonium arboreum* (Crassulaceae)	62
5.	*Eriobotrya japonica* (Rosaceae) Loquat; Níspero del Japón; Bibacier	
6.	*Cicer arietinum* (Leguminosae) Chick pea; Ciuronera, Garbanzo; Zichererbse; Pois chiche	70
7.	*Vicia faba* Broad bean; Haba; Saubohne; Fère	
8.	*Pelargonium* species (Geraniaceae) Geranium; Gitanilla, Geranio, Malva rosa; Geranium; Géranium	
9.	*Vitis vinifera* (Vitaceae) Vine; Llambrusca, Parra, Vid; Weinrebe; Vigne	90
10.	*Hibiscus rosa-sinensis* (Malvaceae) Hibiscus; Cayena; Roseneibisch; Ketmie.	
11.	*Citrullus vulgaris* (Cucurbitaceae) Water melon; Sandia, Patilla; Wassermelone; Pastèque.	
12.	*Punica granata* (Punicaceae) Pomegranate; Magraner, Granado; Granatbaum, Granadier	98
13.	*Olea europaea* (Oleaceae) Olive; Olivera, Aceituno, Olivo; Oelbaum; Oelbaum; Olivier	109
14.	*Ipomoea* sp. (Convolvulaceae) Morning Glory; Campaneta de jardí.	
15.	*Helianthus annuus* (Compositae) Sunflower; Heliantem, Girasol, Corona real; Sonnenblume; Tournesol.	
16.	*Gazania rigens* (Compositae)	
17.	*Cynara scolymus* (Compositae) Globe artichoke; Carxofera, Alcachofera; Artischocke; Artichaut	149
18.	*Scorzonera laciniata* (Compositae)*	152
19.	*Agave americana* (Agavaceae) Century plant; Pita, Filiagulla, Aloe, Magüey; Aloe hauf; Pite	162
20.	*Freesia refracta* (Iridaceae)	164
21.	*Iris pallida* (Iridaceae)	164
22.	*Zanteschia aethiopica* (Araceae) Arum lily; Cala: Aronstab.	

* This is a wild plant, here in error. It is related to *Scorzonera hispanica* (Black salsify; Salsifi negro; Schwarzwurzel; Scorzonère), which has a larger flower, and entire or toothed rather than pinnate leaves. The black tap-root is boiled and eaten as a vegetable.

16
4
19
12
18
14
20
9
7
22
2
5
3
15
13
6
1
21
8
17
10
11

Glumes of central spklt long-ciliate .... *H. murinum* (Ma, Mi, I)
Glumes of central spklt rough or smooth, not long-ciliate.......................
.... *H. marinum* (Ma, Mi, I)

AVENA Oat; Avena; Hafer; Avoine.
Compound, branched infce, spklts long-pedicelled, drooping. Glumes longer than florets, lemmas long-awned from back.

1. Lemma with 2 long bristles at apex.................... *A. barbata* (Ma, Mi)
1. Lemma with 2 short teeth at apex
   2. Spklts 18-25mm...................................... *A. fatua* (?Ma, ?Mi, ?I)
   2. Spklts 25-45mm.. *A. sterilis* (Ma, Mi, I)

A. fatua Wild oat; Ballueca; Wind-Hafer; Avoine folle.

AVENULA
Pnnl; infce branched panicle, spklts erect; lemma with geniculate awn.
Lemma narrowed in upper 1/4-1/3, irregularly toothed at apex ....
.... *A. bromoides* (Ma, Mi, I)
Lemma narrowed to an acute bifid apex .... *A. crassifolia* (I, endemic)

ARRHENATHERUM
A. elatius False oat-grass; Tortero; Glatthafer; Fenasse.
Pnnl, tufted, with loose panicle of spklts, the lower lemma conspicuously geniculate awned from the lower half of the back. (Ma, I)

GAUDINIA
G. fragilis
Softly hairy annl, with slender spike; spklts with conspicuously awned lemmas appressed to rhachis. (Ma, Mi, I)

KOELERIA
K. macrantha
The older names for this and the next species were very confusing. This grass, formerly *Koeleria cristata* (L.) Pers., though widespread in Britain, is very rare in Majorca, and not to be confused with *Lophochloa cristata*, formerly *Koeleria phleoides* Pers. Vill., which is common here.

K. macrantha: small tufted pnnl, with lvs usu. rolled and bristle-like; panicle spike-like, erect, dense, sometimes lobed or interrupted below, 1-19cm long, silvery-purplish and shiny. (Ma)

LOPHOCHLOA

L. cristata

Green annl, spike very variable, from ovate to (more commonly) long or short cylindrical. Lemma 5-veined, shortly bifid at apex, with short, straight awn from near apex. Spklts usu. hairy. It is very common in many types of habitat. (Ma, Mi, I)

TRISETUM

Annls. Spklts in panicles, awn from back of bifid, keeled lemmas. Palea silvery, not enclosed by lemma at maturity.

Awn inserted in upper quarter of back of lemma, straight or curved, greyish .... *T. panicum* (Mi)

Awn inserted about 1/3 way down lemma, twisted and angled, brownish or yellowish .... *T.aureum* (Ma)

LAGURUS

L. ovatus Hare's tail; Cua de ca, Rabillo de conejo ; -; Lagouier

Annl; heads ovoid, dense, very softly woolly. (Ma, Mi, I)

AIRA Hair-grass; Aira; Schmielenhafer; Canche.

Small annl; panicle much branched, erect, wiry, pedicels very slender. Lemma with 2 bristly points and dorsal geniculate awn.

1. Pedicels 2-8 x length spklts
   2. Glumes blunt.. *A. tenorii* (Mi)
   2. Glumes pointed .... *A. elegantissima* (Mi)
1. Pedicels 1-2 x length spklts .... *A. cupaniana* (Ma, Mi, I)

ANTHOXANTHUM

Annl or pnnl, coumarin-scented; lower glume about 1/2 length spike, upper as long as spike, enclosing florets. Lemmas awned.

Annl; culms branched; glumes hairless; awns inconspicuous ....
.... *A. odoratum* (Ma, Mi)

Pnnl; culms unbranched; glumes hairy; awns conspicuous ....
.... *A. aristatum* (Mi)

A. odoratum Sweet vernal-grass; Gram d'olor; Wohlriechendes Ruchgras; Flouve odourante.

HOLCUS

H. lanatus Yorkshire fog; Herba molla, Heno blanco; Honiggras; Houque.

Softly pubescent pnnl; panicle loose, pyramidal in fl., becoming contracted after flowering, soft, often pink; upper floret shortly awned, but awn included within glumes. (Ma, Mi, I)

AVELLINIA

A. michelii

Small annl; stems usu. numerous, puberulent; panicle contracted after flowering. Lower glume 1/4 upper, upper almost as long as florets. Lemma shortly awned from sinus. (Ma, Mi, I)

TRIPLACHNE

T. nitens

Small annl; stems usu. branched below; infce dense cylindrical to ovoid panicle. Glumes both longer than floret, acute, shiny; lemma with geniculate dorsal awn from near base. (Ma, I)

AGROSTIS

Infce lax, pyramidal at flowering time, sometimes contracted after flowering. Spklts small, 1-flowered, glumes enclosing spklt.

Annl; lemma less than half length lowest glume, awn 3mm .... *A. pourretii* (Ma)

Stoloniferous pnnl; lemma at least 2/3 length glumes, usu. unawned .... *A. stolonifera* (Ma, Mi, I)

A. stolonifera Creeping Bent; -; Straussgras; Agrostide.

GASTRIDIUM

G. ventricosum

Small to medium annl; panicles dense, cylindrical or tapering upwards; glumes hard, shining with a long narrow point and bulbous base. Lemmas awned or not. (Ma, Mi, I)

POLYPOGON

Medium tall; panicles dense, often lobed; spikes with 1 floret.

1. Pnnl; heads not conspicuously woolly .... *P. viridis* (Ma, Mi, I)
1. Annl; heads woolly
   2. Glumes notched, shortly ciliate; lemma awned .... *P. monspeliensis* (Ma, Mi, I)
   2. Glumes bifid, long ciliate; lemma unawned .... *P. maritimus* (Ma, Mi, I)

AMMOPHILA

A. arenaria Marram grass; Borró, Barrón; Strandhafer; Oyat.

Stout, rhizomatous, marine pnnl lvs grey-green. Spike stiff, erect, 7-22cm (Ma, Mi, I)

PHLEUM

P. arenarium

Small annl; culms slender; upper lf sheaths rather inflated; panicle very dense, cylindrical or ovoid; spklts 1-fld. (Ma)

ALOPECURUS

A. myosuroides

Tall annl; uppermost lf sheaths rather inflated; panicles spike-like, dense, elongated; glumes united by edges in lower 1/3. Lemma awned from near base, awn exceeding lemma by 4-8mm. (Ma)

PARAPHOLIS

Annl; spklts of 1 floret set in hollow of rhachis.

1. Keel of glumes distinctly winged; spikes usu. straight
   2. Anthers less than 1mm; plant about 10cm, stout, spikes clustered... *P. marginata* (Ma, I)
   2. Anthers at least 2mm; plant usu. more than 10cm, with slender, curved, usu. solitary spikes .... *P. filiformis* (Ma, Mi, I)
1. Keel of glumes not winged; spikes straight to strongly curved.
   3. Spikes often strongly curved; spklts 10-20; anthers less than 1mm... *P. incurva* (Ma, Mi, I)
   3. Spikes straight or slightly curved; spklts 15-30; anthers 2-3mm.. *P. pycnantha* (Ma)

HAINARDIA

H. cylindrica

Small annl; uppermost lf sheath inflated; spike rigid, with dorsally compressed spklts sunk into cavities in rhachis. Lateral spklts with 1 glume, covering cavity except at flowering. (Ma, Mi, I)

PHALARIS Canary grass; Escaiola; Glanzgras; Alpiste.

Panicle dense, spike-like, ovoid or cylindrical; spklts compressed, often with whitish and green stripes; glumes papery, keeled flattened, concealing lemmas, often winged on keel; swollen, firm fertile lemmas each with 2 small sterile lemmas at base.

1. Spklts in groups of 5-7, 1 hermaphrodite, others male or sterile, falling as a group
2. Annl; pedicels hairy; wing of glume with 1 tooth near apex .... *P. paradoxa* (Ma, Mi)
2. Pnnl; pedicels hairless; wing of glume with several small teeth .... *P. coerulescens* (Ma, Mi)
1. Spklts all hermaphrodite (except sometimes basal spklts), persistent
3. Pnnl .... *P. aquatica* (Ma, Mi, I)
3. Annl
4. Glumes with toothed wing .... *P. minor* (Ma, Mi, I)
4. Glumes with entire wing
5. Sterile lemmas at least 1/2 length fertile .... *P. canariensis* (Ma, Mi, introduced)
5. Sterile lemmas not more than 1/5 fertile .... *P. brachystachys* (Ma, Mi, I)

PIPTATHERUM

Pnnl; infce drooping, lax, much branched; spklts of 1 floret with straight awn.

Panicle branches 4 or more at esch node; spklts up to 4mm .... *P. miliaceum* (Ma, Mi, I)

Panicle branches 1-2 at esch node; spklts 8-9mm .... *P. coerulescens* (Ma, Mi, I)

STIPA Feather grass; Espart, Esparto; Federgras; Plumet.

Infce slender, feathery; spklts with 1 very long awned spklt.

1. Awn not geniculate .... *S. bromoides* (Ma)
1. Awn geniculate
2. Lemma membranous, deeply bifid .... *S. tenacissima* (I)
2. Lemma leathery, entire or very shortly bifid
3. Ligule of basal lvs 3.5mm or more, acute; lemma brownish.. *S. offneri* (Ma, Mi, I)
3. Ligule of basal lvs up to 1mm, truncate; mature lemma straw- coloured .... *S. capensis* (Ma, Mi, I)

AMPELODESMOS

A. mauritanica locally: Càrritx, Carrizo, Carcera.

Very large, tufted pnnl, like a small Pampas grass. Numerous long lvs curve outwards from centre of tuft, where arise several much-branched, reed-like infces up to 3m, with short-awned spklts. (Ma, Mi, I)

ARUNDO

A. donax Reed; Canya, Cañavera; Rohr, Flöte; Roseau.

Major landscape features here, these very tall bamboo-like grasses are widely planted as wind-breaks, and for canes for basket making, fencing, roofing and other purposes. (Ma, I)

PHRAGMITES

P. australis Reed; Xisca, Canyet, Carrizo, Cañeta, Cañiza; Schilf; Roseau-á-balais.

Similar to *Arundo*, but much slenderer, up to 3m high. Infce an erect or nodding loose panicle, rhachilla with numerous long hairs. (Ma, Mi, I)

LYGEUM

L. spartum

Tufted pnnl, stems ending in single sheath enclosing 2 or 3 fused spklts; fused part with long white hairs. (I)

AELUROPUS

A. littoralis

Rhizomatous pnnl; infce of flattened spikes, broadside on to rhachis; ligule of long hairs. Usu. maritime (Ma, I)

ERAGROSTIS

Annl; loosely branched infce of numerous elongated, many-fld, awnless spklts; glumes almost equal, keeled.

Lvs with marginal glands.............................. *E. cilianensis* (Ma, Mi, I)
Lvs without marginal glands .... *E. barrelieri* (Ma)

SPOROBOLUS

S. pungens

Creeping, rhizomatous marine pnnl. Lvs rolled upwards, rather stiff, in opposite plaited rows. Panicle small, violet, much-branched, with small crowded spklts. (Ma, Mi, I)

CYNODON

C. dactylon Bermuda grass; Peu de gallina, Agram, Agramen; Hundzahngras; Chiendent, Pied-de-poule.

Mat-forming pnnl, rooting at nodes; numerous erect or geniculate culms, 3-6 spikes clustered at top. Spklts in 2 rows on each spike, appressed to axis. (Ma, Mi, I)

TRAGUS

T. racemosus

Much-branched annl, rooting at nodes; infce spike-like, purplish; glumes unequal, lower minute, papery, upper leathery, with several rows conspicuous hooked bristles. (Ma)

PANICUM Cockspur, Millet; Blat mill, Mil; Hirse; Millet.

Compound branched infce, spklts flattened, upper floret fertile, lower sterile. Lowest glume minute, easily overlooked, so that upper glume and similar lemma are easily mistaken for 2 glumes.

Robust annl, to 1m; infce dense, drooping .... *P. miliaceum* (Ma, I)

Pnnl with creeping, mat-forming stems; infce erect or drooping .... *P. repens* (Ma, Mi, I)

ECHINOCHLOA Cockspur; Cerreig, Pota de gall; Hühnerhirse; Pied-de-coq.

Robust annl; infce of unilateral spike-like racemes branching out from central rhachis.

Racemes up to 3cm; spklts 1.5-3mm in 4 regular rows, unawned .... *E. colonum* (Ma, I)

Racemes up to 10cm, usu. with short branches near base; spklts in 2 to several rows, some awned, usu. more than 3mm ... *E. crus-gallii* (Ma, Mi)

DIGITARIA

D. sanguinalis Crab-grass; Garranchuelo; Fingerhirse; Panic.

Annl; infce of 4-10 spikes, length 5-15cm, all terminal, or in 2 whorls, Spklts in 2 rows, arranged along 1 side of rhachis. (Ma, Mi)

PASPALUM

Annl; infce of usu. 2 terminal spikes. Spklts in one row, of 2 florets, lower sterile, represented by 1 lemma as long as spklt, upper hermaphrodite, with hardened lemma leaving palea exposed.

Upper glume appressed puberulent .... *P. paspalodes* (Ma, Mi, I)

Upper glume glabrous .... *P. vaginatum* (Ma, I)

STENOTAPHRUM

S. secundatum

Rhizomatous pnnl, rooting at nodes. Lvs with short, blunt blades. Infce spike-like, with 2 rows of spklts in cavities in thickened rhachis. Spklts unawned, lower glume ⅓-¼ upper glume. (Ma)

SETARIA Bristle-grass; Carreig, Panissola, Panizo, Panicillos; Borstenhirse; Sétaire.
Infce spike-like, cylindrical, with 1 or more bristles arising from pedicel below spklt.

1. Spklts deciduous on ripening
  2. Upper glume 1/2-2/3 length upper lemma; bristles 5-10 .... *S. pumila* (Ma, Mi)
  2. Upper glume about equalling upper lemma; bristles 1-3
    3. Bristles retrorsely barbed (sticking to clothes) .... *S. verticillata* (Ma, Mi, I)
    3. Bristles antrorsely barbed .... *S. viridis* (Ma, Mi)
1. Spklts not deciduous, only upper glume falling .... *S. italica* (Ma)

PENNISETUM
P. villosum
Tufted small pnnl; infce broadly cylindrical, soft and bristly-hairy, each spike or cluster of spikes surrounded by involucre of long bristles which fall with spklt. (Ma, I)

IMPERATA
I. cylindrica
Robust pnnl; infce 8-15cm, cylindrical, dense silvery. Spklts of 1 floret surrounded by long silky hairs. (Ma, I)

SACCHARUM
S. ravennae
Robust pnnl; infce 30-60cm long, cylindrical, dense and silvery. Larger than *Imperata*, the hairs arise from the axis of the spklt and not below it. (Ma, I)

SORGHUM
S. halepense Johnson grass, Millet; Cañota, Millaca; Sorgho; Herbe de Cuba.
Robust pnnl, lvs broad, with broad white central vein; infce branched, very loose; spklts with shining, ovoid 5mm white fertile floret, and 1 or 2 smaller, purple sterile florets. (Ma, Mi, I)

HYPARRHENIA
H. hirta -; Fenàs de cruca, Fenal, Triguera borde; -; Bardon.
Tufted, usu. robust pnnl; infce branched, with numerous lf-like bracts, and paired purplish spikes at ends of otherwise bare branches. (Ma, Mi, I)

HETEROPOGON

H. contortus

Tufted pnnl; infce a fragile raceme, with paired dissimilar spklts. Awns up to 8cm in upper part infce, similar unawned spklts in lower part. (Ma)

DICHANTHIUM

D. ischaemum

Tufted pnnl, with 2-10 elongated, violet-coloured spikes arising from tip of culm; spklts paired, lower stalkless and awned, the upper stalked and unawned. (Ma)

HEMIARTHRIA

H. altissima

Rhizomatous pnnl; stems procumbent; infce of terminal and axillary spike-like racemes, with one stalked and one unstalked spklt at each node. (BI in FE)

PALMAE

Chamaerops humilis (locally Garballó, Bargulló, Palma d'escombres, Palmito) is the only native palm, a pretty shrublet, with pleated fan-shaped leaves arising from suckering stems. The flowers are small, with six petals, followed by yellow date-like fruit. Sometimes in sheltered places (such as the entrance to the Boquer Valley in Majorca) *Chamaerops* develops a stout blackish trunk and becomes a small tree. (Ma, Mi, I)

Other palms are often planted, and *Phoenix dactylifera* (Date Palm; Palmera, Palma datilera; Dattelpalme; Dattier) actually produces fruit in Majorca. *P. canariensis*, a native of the Canary Islands, lines the elegant boulevards of Palma's sea front. (If mentioning that *P. canariensis* comes from the Canary islands seems to be stating the obvious, just take a look at the endemic species of these islands, which include *Phlomis italica*, *Sibthorpia africana* and *Teucrium asiaticum*.)

Phoenix species have enormous pinnate leaves, *P. dactylifera* with a slender trunk reaching 20m or more, *P. canariensis* with a stout trunk up to about 8m.

ARACEAE

These are juicy rhizomatous perennials. The flowers are minute, on a club-shaped *spadix* which is enclosed in a large, thin bract or *spathe*. The leaves are arrow to inverted heart-shaped, or compound arrow-shaped, and the fruit is a spike of red berries. The best-known species is probably *Zantedeschia aethiopica* (Arum lily), with a bright yellow spadix and a white spathe, often seen in gardens here.

By a curious freak, or perhaps convergent evolution, the spathe of *Arisarum* is very like the perianth of Aristolochia.

KEY

1. Lvs arrow shaped, with secondary arrow-shaped segments arising from lateral angles of primary arrow.. .... *Dracunculus muscivorus* (Ma, Mi)
1. Lvs arrow to inverted heart-shaped, without secondary arrows
  2. Spathe with margins fused at the base, forming a closed tube............ .... *Arisarum vulgare* (Ma, Mi, I)
  2. Spathe with margins overlapping but free at base................. *Arum*

ARUM Lords and ladies, Cuckoo-pint; Sarriasses, Aro; Aronstab; Gouet.

1. Autumn flowering; spathe purple .... *A. pictum* (Ma, Mi, I)
1. Spring flowering; spathe green, with or without purplish markings
  2. Spathe with tip more or less erect; spadix dull purple or brownish-yellow.. *A. maculatum* (?Ma)
  2. Spathe with tip flopping forwards; spadix yellow .... *A. italicum* (Ma, Mi, I)

Several hybrids or possibly other species have been found in Majorca.

LEMNACEAE

The only genus here is *Lemna* (Duckweed; Estrelleta, Lenteja de agua; Wasserlinse; Lenticule d'eau). These are tiny floating plants, with circular green segments which have have minute flowers in pockets on the margin. There are thread-like roots underneath the segments.

Segments often strongly swollen beneath, and reddish or greyish .... *L. gibba* (Ma, I)
Segments flat and green on both sides.................. *L. minor* (Ma, Mi, I)

SPARGANIACEAE

The only species here is *Sparganium erectum* (Branched bur-reed; Bova borda; ästiger Igelkolben; Rubanier). This is a robust hairless perennial, growing in or near water. The leaves are broad linear, and the flowers are in globular clusters. They are borne on a more or less branched inflorescence, male and female in different globules. (Ma, Mi)

TYPHACEAE

There is only one genus in this family, *Typha* (Reedmace, Bulrush; Bova, Ciri, Bayón, Anea; Rohrkolben; Massette, Rauche). The correct English name for this is Reedmace, but it is now more generally known as Bulrush (properly *Scirpus lacustris*), this being blamed on the artist Alma Tadema, who got the species wrong in his painting 'Moses among the Bulrushes'. The tall, densely velvety female spike is probably familiar to most people. The male part of the inflorescence is yellowish, narrow and irregular in outline and lies above the female part.

KEY

Female part of spike at least 10x as long as wide, light brown, usu. with bare rhachis between it and the male part .... *T. domingensis* (Ma, Mi)
Female part of spike up to 6x as long as wide, dark brown, with male part immediately above it...................................... *T. latifolia* (Ma, Mi, I)

CYPERACEAE

This family is often difficult to identify from the small petal-less flowers, which either have no perianth or a perianth reduced to bristles or scales. They are usually arranged in brownish or blackish spikes,each with a solitary bract, or *glume*.

The leaves are very like the leaves of grass but rarely have a ligule. Occasionally the leaves are reduced to a sheath round the stem, which is sometimes triangular in section, especially at the top.

The ovary is superior, and the fruit is often more or less like a very small Spanish chestnut, globular or triangular in section in species with three stigmas, biconvex in species with two, with a beak or point on top. In the sedges the nutlet is wholly enclosed in an outer container formed from the glume (*perigynium*), or closely enfolded by a glume.

KEY

1. Female fls and fr. enclosed in a perigynium .... *Carex*
1. Female fls not enclosed in a perigynium

2. Glumes arranged in 2 opposite rows
   3. Stems lfy; spklts brown, variously loosely clustered .... *Cyperus*
   3. Lvs all basal; infce dense, blackish, ovoid .... *Schoenus nigricans* (Ma, Mi, I)
2. Glumes spirally arranged
   4. All lvs reduced to sheaths or absent
      5. Infce a solitary terminal spike.......... *Eleocharis palustris* (Ma, Mi)
      5. Infce of several spikes, usu. branched .... *Scirpus*
   4. At least some lvs with a green blade
      6. Stems pubescent at nodes and apex; glumes pubescent.. *Fuirena pubescens* (Ma)
      6. Stems entirely or almost hairless; glumes hairless
         7. Infce terminal and in axils of lf-like bracts; lvs and bracts sharply serrate.................................. *Cladium mariscus* (Ma, Mi, I)
         7. Infce terminal only; lvs and bracts not sharply serrate, sometimes rough on margins.. *Scirpus*

Cladium mariscus Sedge; Cesquera, Jonc moresc; Schneide; Marisque.
Schoenus nigricans Bog-rush; -;Kopfried; Choin.

SCIRPUS

1. All lvs reduced to sheaths, or absent
   2. Lower bract as long as infce or longer................ *S. littoralis* (Ma, I)
   2. Lower bract shorter than mature infce.......... *S. lacustris* (Ma, Mi, I)
1. At least some lvs with a green blade
   3. Slender annl, rarely exceeding 15cm; lvs 1-2, thread-like; spikes sessile, 1-3, rarely exceeding 5mm
      4. Bract longer than infce; nut ribbed, shiny .... *S. setaceus* (Ma)
      4. Bract not or hardly longer than infce; nut smooth, not shiny .... *S. cernuus* (Ma, Mi, I)
   3. Stout pnnl, 30cm or more; lvs numerous, broad (sometimes submerged); spikes numerous, sessile or at the ends of terminal branches, usu. exceeding 5mm
      5. Spikes globular.................................. *S. holoschoenus* (Ma, Mi)
      5. Spikes egg-shaped
         6. Stems triangular in section.................. *S. maritimus* (Ma, Mi, I)
         6. Stems circular in section........................ *S. lacustris* (Ma, Mi, I)

CYPERUS

1. Infce a sessile head, or cluster of sessile spklts
   2. Annl; glumes 1mm, acute................................ *C. fuscus* (Ma, Mi)

2. Pnnl; glumes 2.5mm or more, obtuse
   3. Infce a dense, sessile, hemispherical to globose head; glumes 5-8mm, mucronate .... *C. capitatus* (Ma, I)
   3. Infce a cluster of 1-40 sessile spklts; glumes 2-2.5mm, not mucronate .... *C. laevigatus* (Ma, Mi, I)
1. Infce a simple or compound umbel
   4. Annl; longest branches up to 1cm...................... *C. fuscus* (Ma, Mi)
   4. Pnnl; branches up to 6cm or more, longest exceeding 1cm
      5. Longest rays more than 10cm......................... *C. longus* (Ma, Mi)
      5. Rays not exceeding 10cm
         6. Rhizomes ca 1mm wide; spklts brownish... *C. rotundus* (Ma, Mi, I)
         6. Rhizomes large, globular; spklts golden.......... *C. esculentus* (Ma)

CAREX Sedge; Jonça, Junco; Segge; Laîche.

1. Spikes all more or less similar
   2. Stigmas 3
      3. Leaves exceeding stems, 1.5-2mm wide; lowest spike long-stalked, arising from near base of stem .... *C. oedipostyla* (Mi)
      3. Leaves about equalling stem; lowest spklt short-stalked, arising from about middle stem .... *C. distachya* (Ma, Mi, I)
   2. Stigmas 2
      4. Plant densely tufted, without creeping rhizomes
         5. Stems less than 2mm thick; upper spikes crowded, lower remote; fr. with feint veins only at base .... *C. divulsa* (Ma, Mi)
         5. Stems more than 2mm thick; spikes crowded at top of stem; fr. densely veined
            6. Stems very sharply angled or winged, faces slightly concave; infce reddish-brown.. *C. vulpina* (Ma)
            6. Stems triangular in section, not winged, faces flat; infce yellowish-green or light brown, dark brown in fr..... *C. otrubae* (Ma, Mi)
      4. Plant with creeping rhizomes, hardly tufted ...... *C. divisa* (Ma, Mi)
1. Spikes dissimilar, terminal usu. entirely male, lower entirely female
   7. Stigmas 2; fr. usu. biconvex or flat on one side and convex on the other .... *C. acuta* (Ma)
   7. Stigmas 3; fr. usu. smooth or triangular in section
      8. Female spikes 40mm or less
         9. Stems filiform; lvs up to 1mm wide; female spikes 4-6mm .... *C. rorulenta* (Ma, endemic)
         9. Stems not filiform; lvs 1-5mm wide; female spikes 8-40mm
            10. Male spike solitary, 5-10mm; lower female spikes long-stalked, from base of stem .... *C. depressa* (?BI in FE)

10. Male spikes solitary or not, 10mm or more; lower female spikes sessile or short-pedunculate, not from base of stem
 11. Leaves glaucous; fr. minutely papillose, beak entire .... *C. flacca* (Ma, Mi, I)
 11. Leaves green; fr. not papillose; beak bifid or notched
  12. Male spikes 1 or 2, 15-30mm; fr. conspicuously many-ribbed, beak bifid, rough .... *C. distans* (Ma, Mi, I)
  12. Male spike 1, 10-15mm; fr. weakly ribbed, beak slightly notched, smooth .... *C. extensa* (Ma, Mi, I)
8. Female spikes 40-120mm .... *C. hispida* (Ma, Mi, I)

ORCHIDACEAE

Europeaen orchids lack something of the style and glamour of the exotic species sold in florists' shops, and their fans, showing slides to non-botanist friends, may have been hurt by the cool reception. But the real difference is simply one of size, and anyone prepared to take time to examine almost any European orchid with a lens will be struck by the beauty and complexity of the flower.

The perianth, as in Liliaceae, is composed of two alternating whorls of three segments. The inner whorl usually has two small lateral segments and a much enlarged *labellum* below. There is often a spur at the base of the labellum. The outer whorl has three similar segments. The colouring of the whorls is very variable, and either whorl may be coloured (petaloid) or green (sepaloid).

The anthers and stigmas are borne on a special structure called the *column*. The (here) single stamen is borne near or at the tip of the column, where the pollen from each of the two anther cells is aggregated into *pollinia*, which are globular waxy masses which stick to a pollinating insect.

The stigmas are three, one of which is (here) sterile (sometimes absent), and the other two usually represented by a stalkless sticky surface for receiving the pollinia. The sterile stigma sometimes forms a projection between the anthers and the fertile stigmas to reduce the chance of self-fertilisation, it is then called the *rostellum*.

In the following description 'inner perianth segments' refers to the upper two segments – where necessary the labellum is described separately.

KEY

1. Plant lacking chlorophyll, leaves reduced to scales, not green
2. Whole plant violet .... *Limodorum abortivum* (Ma, Mi)

2. Whole plant pale yellowish brown .... *Neottia nidus-avis* (Ma)
1. Plant with green leaves
3. Spur absent
4. Both perianth whorls (including labellum) more or less uniformly pink or white, sometimes with yellowish or darker markings
5. Leaves in basal rosette lateral to flowering stem (small pale green bracts present on stem); flowers 4-5mm, white, spirally arranged .... *Spiranthes spiralis* (Ma, Mi)
5. Large stem leaves present; fls 15mm or more .... *Cephalanthera*
4. Perianth segments not uniformly pink or white
6. All perianth segments except labellum connivent to form a hood over column
7. Outer perianth segments free; labellum shaped like a man, greenish or yellowish with darker markings .... *Aceras anthropophorum* (Ma, I)
7. Outer perianth segments fused at the base; labellum with pink, purplish or whitish tongue-like central portion, and two darker basal lobes .... *Serapias*
6. Perianth segments spreading, not connivent
8. Inner perianth segments slightly smaller than outer; labellum not hairy, dark or insect-like.. *Epipactis*
8. Inner perianth segments much smaller than outer; labellum hairy, dark, and often insect-like .... *Ophrys*
3. Spur present
9. Spur more than 10mm long
10. Fls uniformly greenish-white; labellum narrow, elongated and entire.. *Platanthera bifolia* (?BI)
10. Fls not uniformly greenish-white; labellum usu. 3 to 5-lobed or dentate, if entire, then broad and crenulate
11. Spur very slender, 11mm or more, less than 1mm diam., curving downwards
12. Labellum shallowly 3-lobed; spur almost twice length of ovary .... *Gymnadenia conopsea* (Ma)
12. Labellum deeply 3-lobed; spur equalling or exceeding ovary......................... *Anacamptis pyramidalis* (Ma, Mi, I)
11. Spur more than 1mm wide in parts at least, usu. straight
13. Bracts all membranous .... *Orchis*
13. At least lower bracts lk-like .... *Dactylorhiza*
9. Spur less than 10mm long
14. Fls 7mm .... *Neotinea maculata* (Ma, Mi)
14. Fls 10mm or more

15. Bracts herbaceous, green .... *Dactylorhiza*
15. Bracts membranous
16. Plant very robust; central lobe of labellum deeply divided; (flowering Dec.-March)............. *Barlia robertiana* (Ma, Mi)
16. Plant small or tall and slender; central lobe of labellum not deeply divided .... *Orchis*

Neottia nidus-avis Birds' nest orchid; Nido de ave, Nido de pájaro.
Aceras anthropophorum Man orchid; Flor de l'home penjat; Fratzenorchis; Homme pendu.

EPIPACTIS

1. At least outer perianth segments greenish inside
2. Lvs up to 2.5cm long, shorter than internodes .... *E. microphylla* (Ma)
2. Lvs much longer, exceeding internodes .... *E. helleborine* (Ma)
1. Outer perianth segments brownish-pink inside .... *E. palustris* (?BI in FE)

CEPHALANTHERA

1. Flowers pink .... *C. rubra* (?Ma)
1. Flowers white
2. Bracts longer than ovary; outer perianth segments blunt .... *C. damasonium* (Ma)
2. Bracts shorter than ovary; outer perianth segments acute .... *C. longifolia* (Ma)

DACTYLORHIZA

Fls yellow or yellowish; spur as long as ovary..... *D. sulphurea* (?BI in FE)
Fls pinkish; spur shorter than ovary................. *D. maculata* (?BI in FE)

ORCHIS Orchid; Botons de gos; Knabenkraut; Orchidée.

1. Lateral outer perianth segments spreading or reflexed
2. Spur as long as ovary or longer .... *O. mascula* (Ma)
2. Spur shorter than ovary
3. Labellum 3-lobed
4. Outer perianth segments greenish inside with purple spots.. *O. patens* (?BI)
4. Outer perianth segments pinkish purple inside and out .... *O. laxiflora* (Ma)
3. Labellum oval with marginal teeth, unlobed.......... *O. saccata* (Ma)
1. All perianth segments except labellum connivent to form a hood

5. Labellum unlobed .... *O. papilionacea* (?Ma)
5. Labellum 3-lobed
  6. Spur horizontal or curving upwards; outer perianth segments green veined
    7. Spur about equalling labellum, or slightly longer, not thickened at tip .... *O. morio* (BI in FE)
    7. Spur much longer than labellum, thickened at tip .... *O. longicornu* (Ma, Mi)
  6. Spur directed downwards; outer perianth segments not green-veined
    8. Labellum with long pointed lobes (naked man) ...... *O. italica* (Ma)
    8. Lobes of labellum not long pointed
      9. Labellum unspotted or with spots around central area; middle lobe narrow at margin, with 1-3 teeth .. *O. coriophora* (Ma, Mi, I)
      9. Labellum with spots mainly on lobes; middle lobe broad, with many teeth at margin .... *O. tridentata* (Ma, Mi)

SERAPIAS Tongue orchid; -;Stendelwurz; Sérapias.
(The labellum here, apparently entire, is actually 3-lobed. The two lateral lobes are curved upwards and partly hidden inside hooded part of flower. 'Tongue' here refers to the central lobe of the labellum.)

1. Tongue narrowed at base
  2. Bracts longer than fls.................................... *S. vomeracea* (Mi)
  2. Bracts equalling or shorter than fls
    3. Fls 25-40mm; tongue hanging vertically......... *S. lingua* (Ma, Mi, I)
    3. Fls 15-20mm; tongue reflexed .... *S. parviflora* (Ma, Mi, I)
1. Tongue not narrowed at base............................ *S. cordigera* (Mi)

OPHRYS Bee Orchids; Sabatetes; Ragwurz, Kerfstendel; Ophrys.
(There are very many subspecies and varieties of this genus in Europe, as well as many hybrids. It is stressed that, as elsewhere in the book, hybrids are omitted, and the key applies only to subspecies and varieties known to occur here.)

1. Outer perianth segments pink (sometimes very pale whitish- or greenish-pink)
  2. Centre of lower margin labellum with conspicuous, forwards pointing yellow or greenish knob or pointed appendage
    3. Inner perianth segments more than half length outer, linear; labellum long and narrow, usu. forwards directed, sometimes nearly horizontal.. *O. bertolonii* (Ma, Mi, I)

3. Inner perianth segments less than half length outer, usu. triangular; labellum usu. directed downwards,
   4. Labellum with conspicuous broad yellow or greenish margin .... *O. tenthredinifera* (Ma, Mi, I)
   4. Labellum brown or blackish to margins, or with narrow green or yellow border, wider near tip
      5. Lower part of labellum broad, shallowly notched at centre .... *O. fuciflora* (Mi)
      5. Lower part of labellum more or less convergent to central point... .... *O. scolopax* (BI in FE)
2. Centre of labellum with backwards pointing central appendage .... *O. apifera* (Ma, Mi, I)

1. Outer perianth segments green
   6. Labellum with broad, lobed yellow or greenish margin .... *O. lutea* (Ma, Mi, I)
   6. Labellum without yellow or greenish margin
      7. Fls less than 1cm; labellum not much longer than outer perianth segments... *O. bombyliflora* (Ma, Mi, I)
      7. Fls more than 1cm
         8. Labellum conspicuously shaggy fringed
            9. Labellum with large central, inverted heart-shaped, bluish (rarely brownish), shiny mirror-like portion.. *O. speculum* (Ma, Mi, I)
            9. Labellum with more or less H-shaped bluish markings (lower part of H and cross connection more constant than upper part) .... *O. sphegodes* (Ma)
         8. Labellum not conspicuously shaggy fringed, bluish markings not inverted heart-shaped or H-shaped .... *O. fusca* (Ma, Mi, I)

O. speculum Mirror orchid ; -; Spiegel-Ragwurz; Ophrys miroir.
O. fusca Sombre bee-orchid ; -; Ophrys sombre.
O. fuciflora Spider orchid ; -; Hummel-Ragwurz. *O. apifera* Bee-orchid; Mosques d'ase; Flor de l'abella, Flor de la abeja; Bienen-Ragwurz; Ophrys abeille.
O. bombyliflora Bumble-bee orchid; Mosques petites.

Explanatory figures to show basic plant structures used in keys

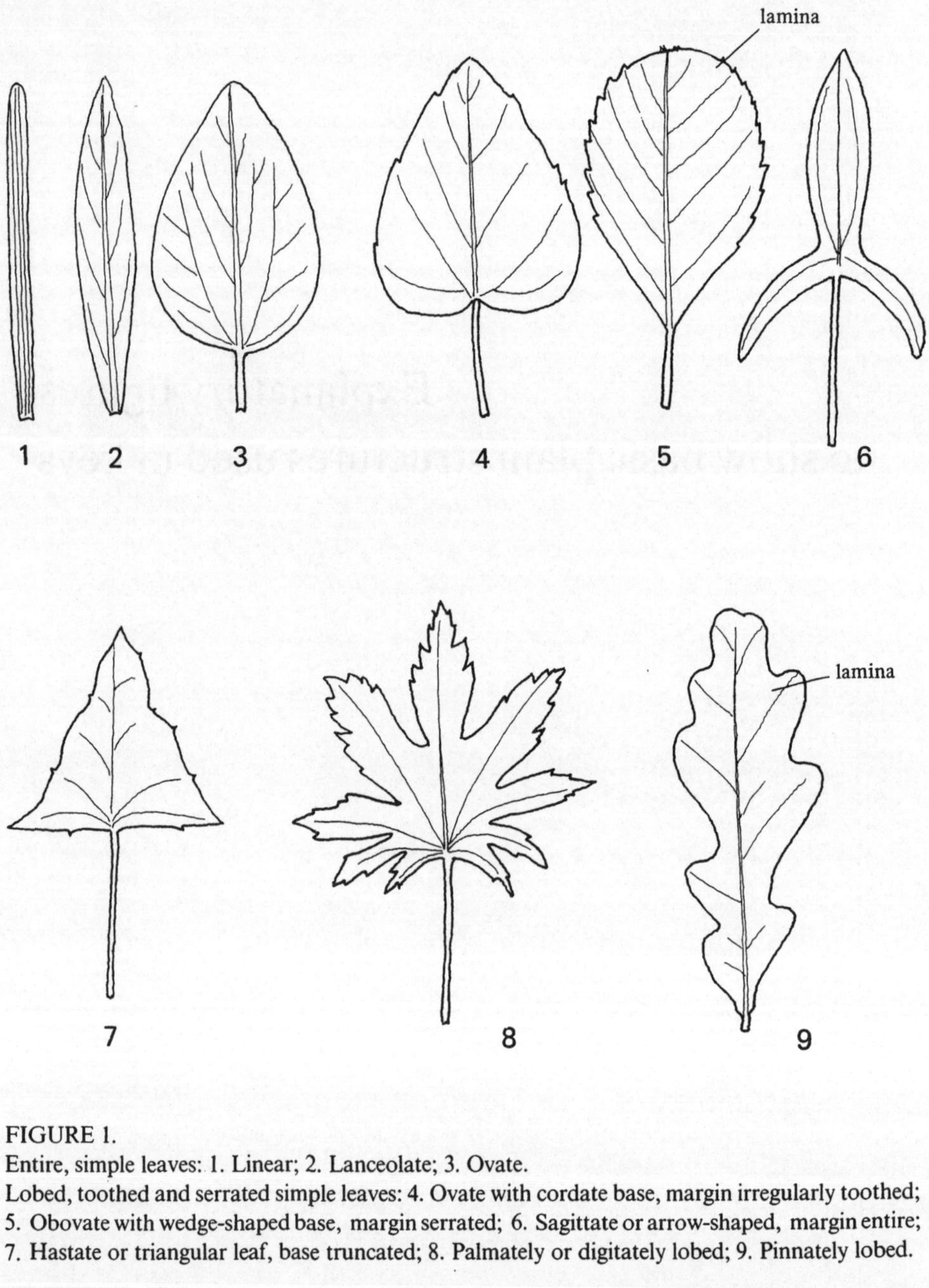

FIGURE 1.
Entire, simple leaves: 1. Linear; 2. Lanceolate; 3. Ovate.
Lobed, toothed and serrated simple leaves: 4. Ovate with cordate base, margin irregularly toothed; 5. Obovate with wedge-shaped base, margin serrated; 6. Sagittate or arrow-shaped, margin entire; 7. Hastate or triangular leaf, base truncated; 8. Palmately or digitately lobed; 9. Pinnately lobed.

Compound leaves (see opposite page): 10. Pinnate with terminal leaflet (or pinna); 11. Pinnate with terminal tendril, base stipulate; 12. Bipinnate; 13. Palmate or digitate; 14. Pinnately trifoliate; 15. Palmately trifoliate.
Arrangement of leaves on stem: 16. Alternate leaves; 17. Opposite leaves, bases connate; 18. Spirally arranged; 19. Sheathing; 20. Sheathing, with ligule at junction of blade with sheath.

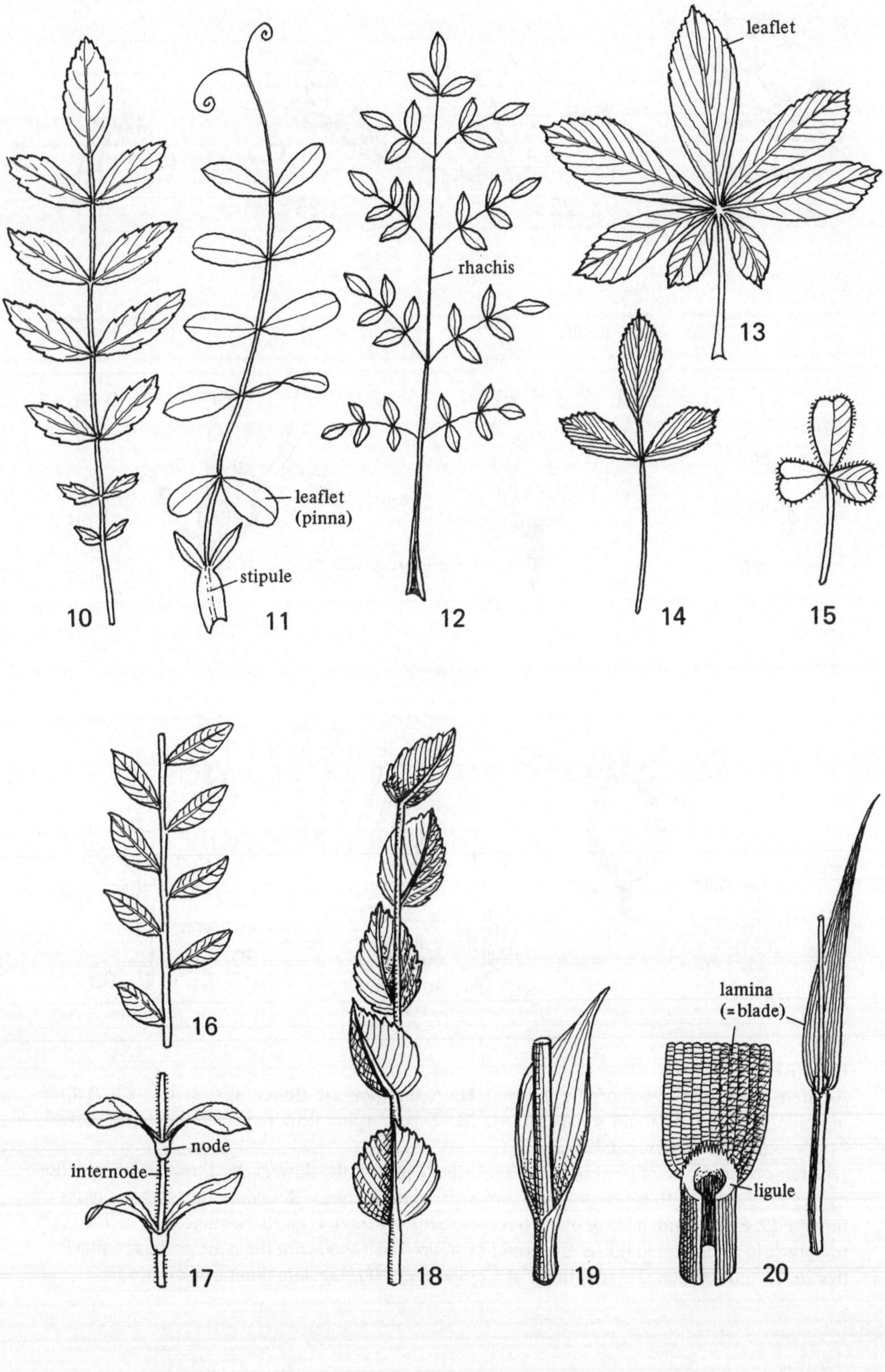
leaflet
rhachis
leaflet
(pinna)
stipule
10
11
12
13
14
15
16
node
internode
17
18
19
lamina
(=blade)
ligule
20

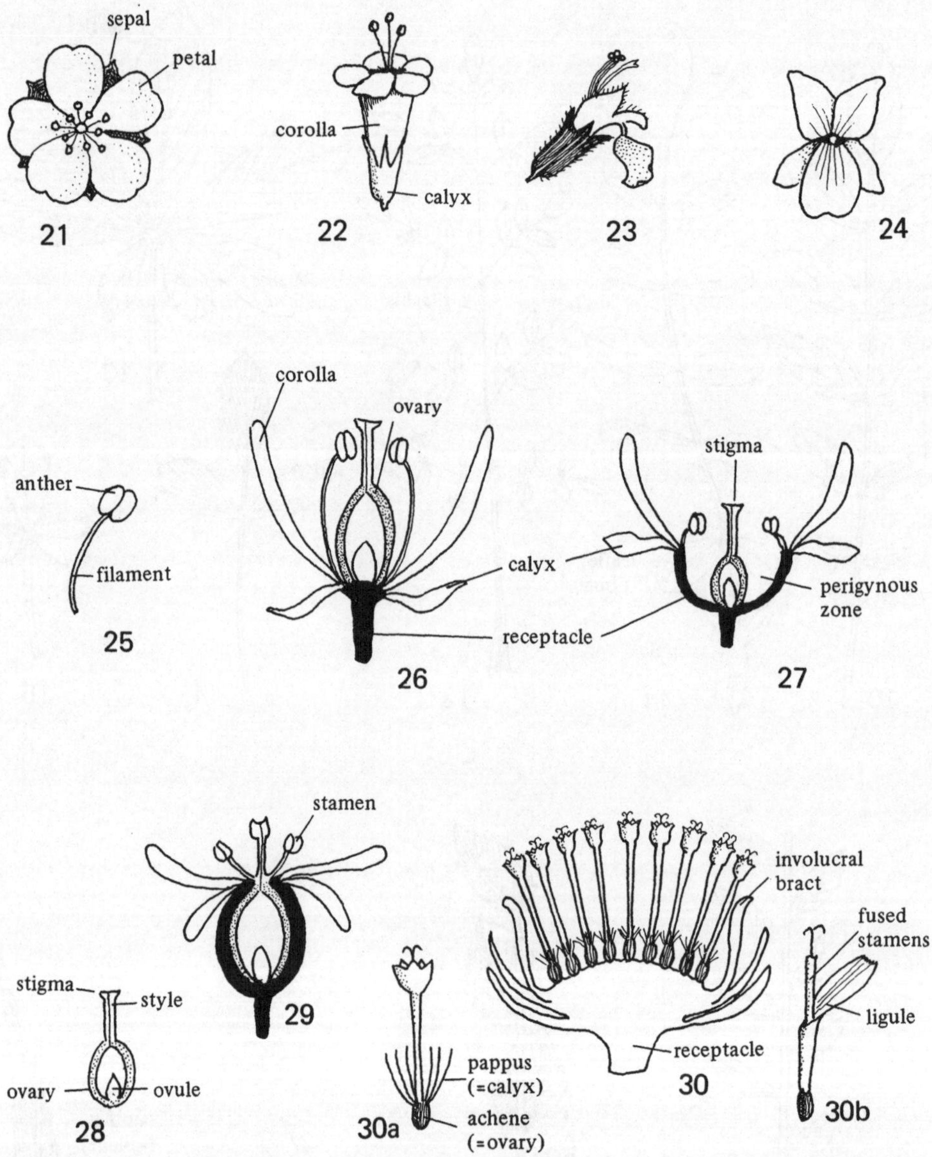

FIGURE 2.
Actinomorphic and zygomorphic flowers: 21. Actinomorphic flower, petals free; 22. Actinomorphic flower, petals and sepals united; 23. Zygomorphic flower, petals and sepals united; 24. Zygomorphic flower, petals free.
Diagrams of structure of flowers: 25. Stamen (male part of the flower); 26. Flower with superior ovary; 27. Flower with superior ovary and a perigynous zone; 28. Gynoecium (female parts of flower); 29. Flower with inferior ovary (receptacle fused with ovary); 30. Arrangement of florets on receptacle in a compound flower (It should be noticed that this is not the same as the receptacle in figs 26, 27 and 29); 30a. Tubular floret of Compositae; 30b. Ligulate floret of Compositae.

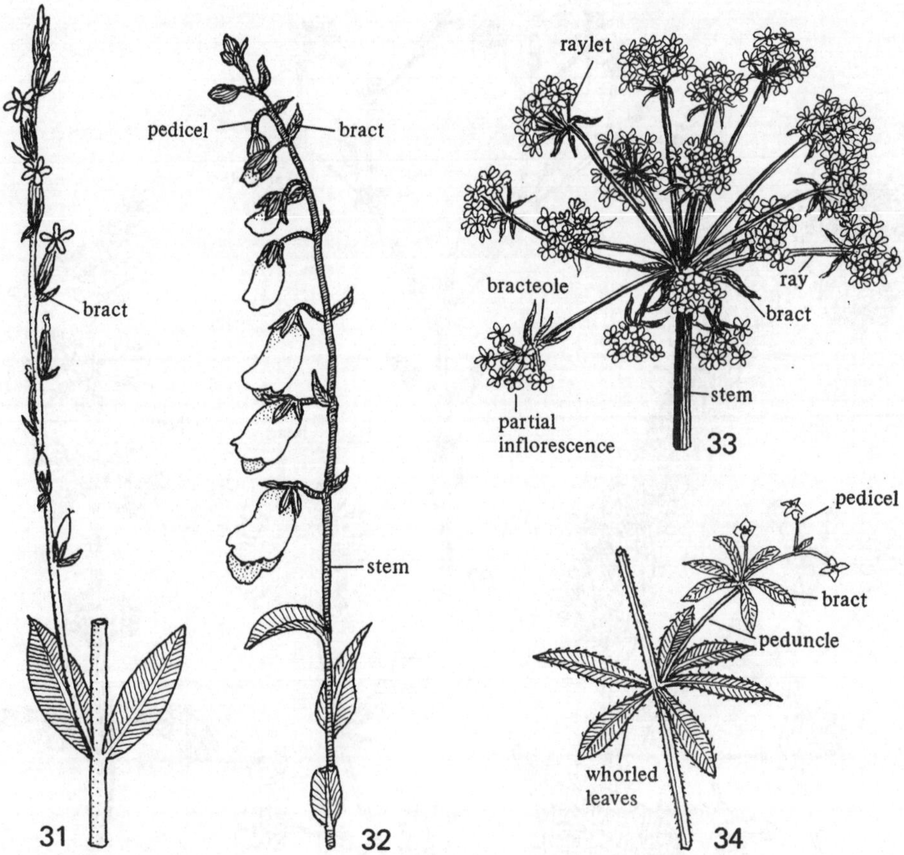

FIGURE 2 (continued).
Inflorescences: 31. Axillary spike; 32. Raceme; 33. Umbel (this is actually a compound umbel, each ray having a small umbel peripherally); 34. Partial inflorescence.

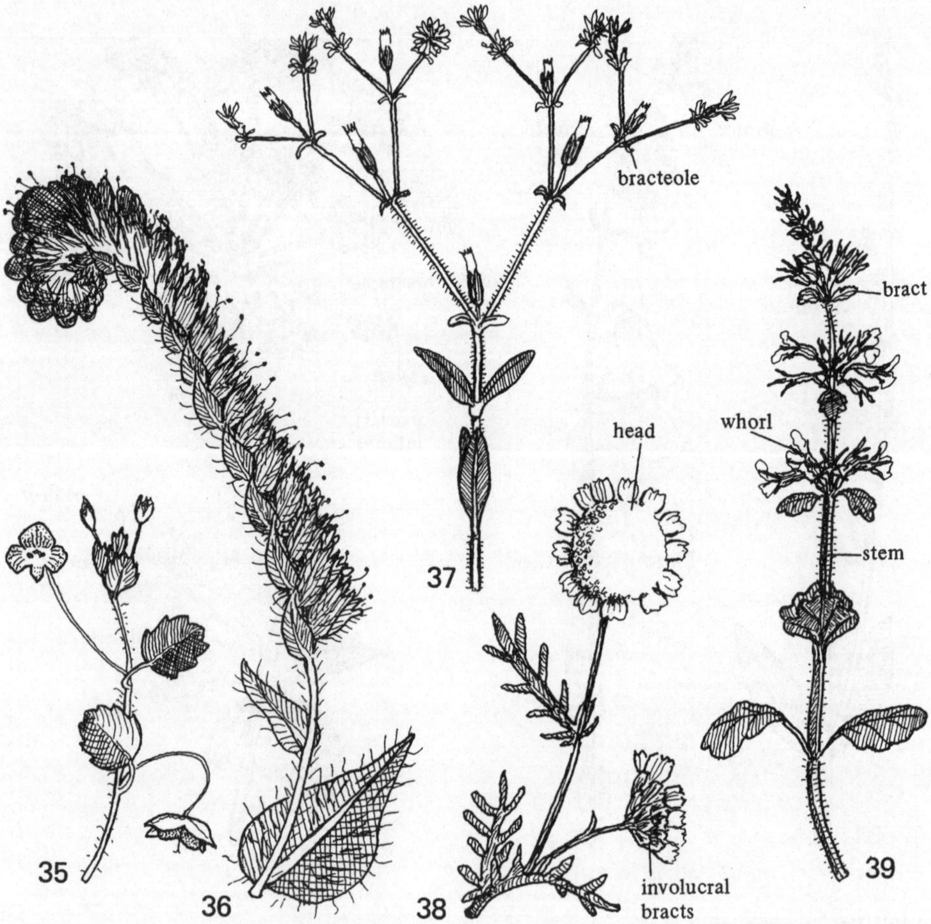

FIGURE 2 (continued).
Inflorescences: 35. Solitary axillary flowers. The distinction between a plant with solitary axillary flowers and one with a bracteate spike or raceme is often a fine one, depending whether the flowers are in the axils of leaf-like leaves (solitary axillary flowers), or less leafy bracts, as in figs 31 and 32; 36. Scorpioidal cyme. This uncoils so that the flowers are at the top, with dead flowers and developing fruit on the straight part. The buds are in the coil; 37. Dichasial cyme. New buds are produced on either side of the flower at the tip; 38. Capitate inflorescence. Flowers in heads, either (as here) in a compound flower (fig. 30), or any other arrangement with the flowers closely packed in a more or less globular form with a stalk; 39. Inflorescence with whorled flowers. Strictly this means that the flowers are sessile or on pedicels arising in a circle from the main stem; but is used here, as elsewhere, to include arrangements with shortly pedunculate branches in the 'whorls'.

Glossary

Text figures are placed on the right side of the Glossary.

See also Figures 1 and 2 at end of glossary.

Achene: a small, dry one-seeded fruit. Fig. 2:30a.

Actinomorphic: radially symmetrical, like a wheel. Fig. 2:21.

Alternate: (of leaves): strictly with leaves first on one side then the other side of the stem, in two rows; used more generally, as here, to mean not in opposite pairs. Fig. 1:16.

Annual: a plant which germinates, goes through its complete life-cycle and dies in one year.

Anther: the pollen bearing organ of a plant. Fig. 2:25.

Appressed: pressed against (usually of hairs pressed against a stem, not sticking out horizontally from stem).

Arista: a bristly projection. Text fig. 1.

1

Aristate: (adj.) with an arista.

Arrow-shaped: Fig. 1:6.

Auricle: ear-like projection at the base of a leaf (often clasping stem). Text fig. 2.

2

Awn: stiff, bristle-like projection (esp. in grasses).

Axil: the angle between a leaf or a bract and the stem.

Axillary: in the axil(s). Fig. 2:35.

Basal: at the base or lower end.

Berry: a fleshy fruit with one or more seeds in the pulp, not inside a stony or cartilaginous endocarp.

Biennial: a plant which germinates in the first year, and flowers, produces fruit, and dies in the second year.

Bifid: split into two.

Bipinnate: with the pinnae themselves pinnate.

Bract: a scale or small leaf-like structure. In plants with a spike or racemose inflorescence there may be a gradual transition up the stem from typical leaves to bracts. Fig. 2:31, 32, 33.

Bracteole: a bract on a peduncle or pedicel, not the main stem. Fig. 2:31,32.

Bristle: a stiff hair.

Bulbil: small bulbs produced vegetatively, often above ground, in some genera especially *Allium*.

Caducous: transient, falling soon after development.

Calyx: green or dull coloured perianth whorl, the only or outer whorl. Fig. 2:22,26,27,29.

Capsule: a dry, usually thin-walled, fruit formed from 2 or more fused carpels, splitting open when ripe.

Carpel: a single one of the units of which a gynoecium is made up (there is often only one).

Carpophore: axis of flower between calyx attachment and base of ovary (especially in *Silene*). Text fig. 3.

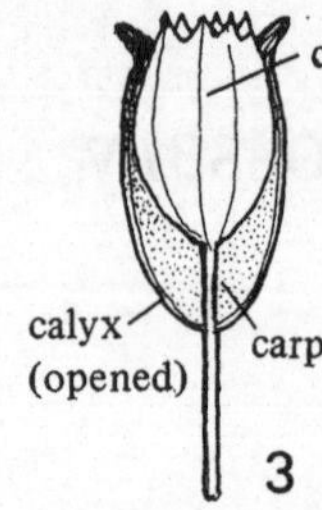

3

Cladode: modified branch of stem which appears leaf-like, especially in Liliaceae. Text fig. 4.

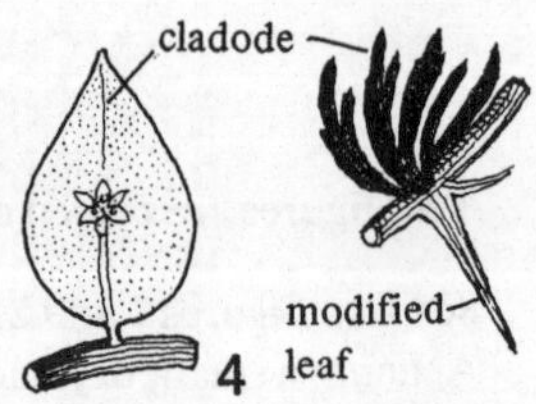

4

Claw: narrowed inner end of a petal. Text fig. 5.

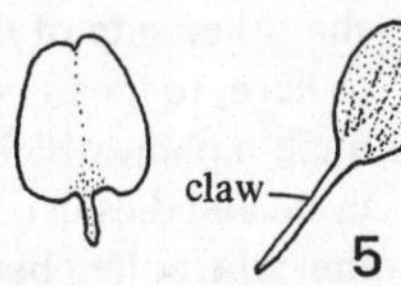

5

Cleistogamous: fertilizing itself without opening.

Clone: a group of cells, or, more loosely, individuals which have arisen by non-sexual reproduction, and are therefore genetically identical.

Cluster: a branched inflorescence with several flowers close together.

Compound: an inflorescence in which several or many flowers, inflorescence or florets, are set side by side upon a variously shaped receptacle, like beads on a brooch (e.g. a daisy). Fig. 2:30.

Compound leaf: a leaf formed of several leaflets (like small leaves) joined to a main axis, or rhachis, all falling together. Fig. 1:10-17.

Connate: fused together.

Cordate: heart-shaped, notch downwards. Fig. 1:4.

Corolla: brightly coloured whorl of perianth, only or inner whorl, composed of free or united petals. Fig. 2:22,26.

Corona: perianth-like whorl in the centre of a flower attached to inside of the petals (e.g. the tubular central part of a daffodil, or something much smaller, especially in *Silene*). Text fig. 6.

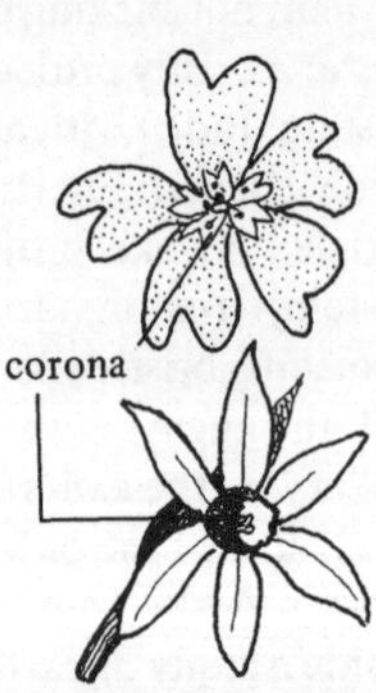

6

Cyme: a branched inflorescence, usually obconical in general shape, in which new growing points are continually produced by division of the terminal parts of the inflorescence, so that the newer flowers are at the periphery. Fig. 2:36,37.

Deciduous: falling off, especially of leaves in autumn. (A deciduous tree is one that loses all its leaves in autumn).

Decumbent: lying on the ground, with growing tips turning upwards.

Decurrent: running down (especially of prolongation downwards of blade of sessile leaf to form a wing on the stem). Text fig. 7.

7

Dehiscent: breaking open when ripe (of fruit).

Dichasial: of a cyme which branches repeatedly into two approximately equal branches, often with a flower at the point of each division. Fig. 2:37.

Dichotomous: dividing into two equal branches.

Dioecious: with male and female flowers on separate plants.

Drupe: a fleshy fruit containing one or more seeds, each surrounded by a stony endocarp (e.g. a plum).

Endocarp: the lining of the cavity containing the seed.
Entire: not toothed, notched or cut. Fig. 1:1-3.
Epicalyx: a calyx-like structure outside the calyx. Text fig. 8.
Evergreen: not shedding all leaves simultaneously.
Exserted: sticking out (not enclosed). e.g. of stamens in a tubular corolla. Text fig. 9.
Fertile: capable of producing viable seed or pollen.
Filament: the stalk of an anther. Fig. 2:25.
Floret: a small flower, usu. part of a compound infce. Fig. 2:30a,30b.
Gland: a small globular vesicle containing a fluid, often oily or sticky or scented (pleasantly or unpleasantly), protruding from or embedded in part of the plant. Text fig. 10.
Glandular hair: a gland with a stalk. Text fig. 10.
Glaucous: a bluish-green colour, characteristic of some plants.
Gynobasic: arising from near the lower end of the ovary (of a style, especially in Labiatae). Text fig. 11.
Gynoecium: the female reproductive parts of a plant, consisting of carpels with a stigma (with or without a style). Fig. 2:28.
Hastate: spear-shaped. Fig. 1:7.
Head: an inflorescence of many closely set small flowers or florets, especially (but not only) of a compound inflorescence. Fig. 2:38.
Herb: a plant without woody parts.
Herbaceous: Green and soft (not woody).
Hedgehog plant: a plant like a hedgehog – a small, dense, prickly mound.
Hermaphrodite: with fertile male and female parts.
Hispid: with rough bristles.
Imbricate: overlapping, like tiles on a roof. Text fig. 12.
Indusium: a membranous piece of tissue covering the sporangium (in ferns). Text fig. 13.
Inferior ovary: an ovary with the petals and calyx inserted at the outer end (like an apple, the fruit developing beneath the flower). Fig. 2:29.
Inflorescence: the flowering part of a plant, with branches, flowers and often bracts. Fig. 2:31-39.
Internode: interval between nodes. Fig. 1:17.
Involucre: calyx-like structure surrounding head of a compound flower. Fig. 2:30,38.
Involucral: bracts. Sepal-like bracts forming the involucre. Fig. 2:30,38.
Keel: the sharp angle where two surfaces meet (like the keel of a boat). In Leguminosae the term includes the surfaces as well, the boat-like structure formed by fusion of the two lower petals by their adjacent margin. Text fig. 14.
Lamina: the soft blade of a leaf, not the stalk and central vein. Fig. 1:5,9,20.
Lanceolate: lance-shaped. Fig. 1:2.

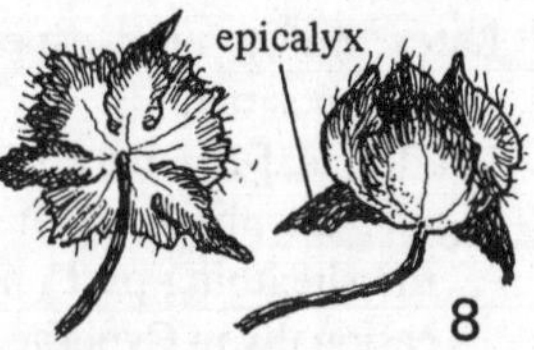

8

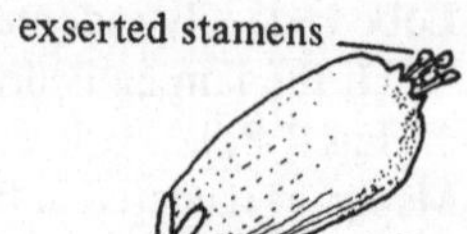

9

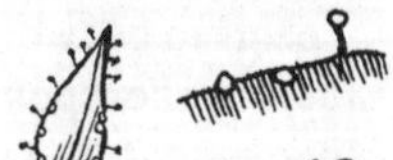
10

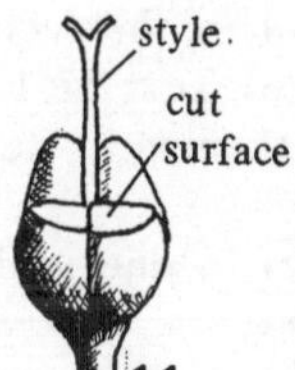

11

12

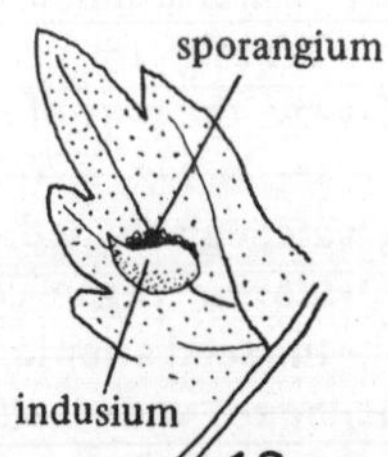

13

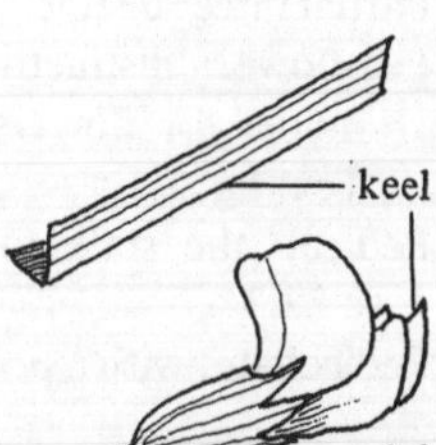

14

Latex: milky juice, especially of *Euphorbia* and some genera of Compositae.

Leaflet: leaf-like part of a compound leaf. Fig. 1:11,13.

Ligule: membranous or hairy structure at the junction between the sheathing part and the blade of the leaf of a grass (or occasionally in Cyperaceae). Fig. 1:20.

Linear: long and narrow with parallel sides. Fig. 1:1.

Lobed: deeply indented, (in a leaf) with the lamina continuous. (If the lamina is divided to the rhachis the leaf is compound). Fig. 1:8,9.

Mealy: as if coated in flour.

Membranous: thin and usually translucent, not green.

Medifixed: (of hairs) joined by the centre, so that one hair appears to be two. Text fig. 15.

Mucro: a short point. Text fig. 16.

Mucronate: (adj.) with a mucro.

Nectary: an organ producing nectar (especially in Ranunculaceae, where the nectary is petal-like, with a small vesicle near the base). Text fig. 17.

Node: thickening in stem where one or more leaves arise. Fig. 1:17.

Nut, nutlet: a one-seeded, indehiscent fruit with a hard outer covering.

Ob-: a prefix meaning upside-down, or inverted; or wider at the top. In the case of the word cordate this can be confusing, because cordate starts as a heart upside down, so obcordate puts it the right way up again. Where this leads to ambiguities in the key 'notch upwards' or 'notch downwards' has been added.

Ochrea: a membranous sheath or partial sheath formed from fused pl.

Opposite: applied to leaves, in pairs on opposite sides of stem. Text fig. 18.

Ovary: the part of the flower containing the ovules, and later the seeds. Fig. 2:26-29.

Ovate: egg-shaped. Fig. 1:3.

Ovule: an organ containing the female gamete, and becoming the seed after fertilization. Fig. 2:28.

Palmate: with leaflets radiating from a point, like fingers on a hand (=digitate). Fig. 1:13.

Partial infce: distinct part of a branched infce. Fig. 2:33.

Pea-flower: distinctive flower of *Pisum sativum*, here used as model for flowers of Leguminosae sub-family Papilionatae. Text fig. 19.

Pedicel: the stalk of a single flower in an inflorescence. Fig. 2:32,34.

Pedicellate: with a pedicel.

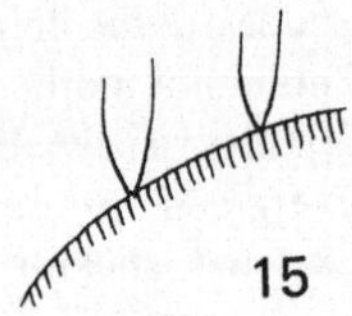
15

16

17

18

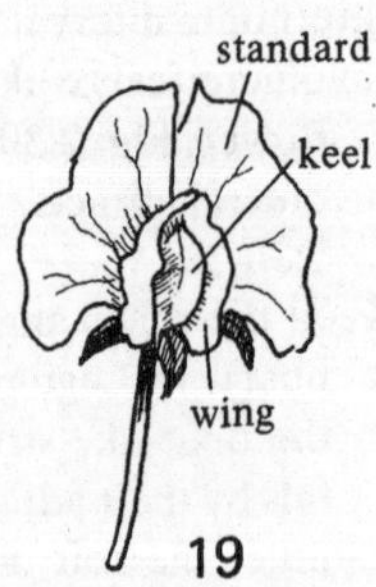

19

Peduncle: the stalk of an inflorescence or partial inflorescence (from which the pedicels branch). Fig. 2:34.

Pedunculate: with a peduncle.

Perennial: a plant that continues to flower and set seed for more than one year, usually recognized by the presence of non-flowering shoots among the flowering ones – an annual or biennial plant flowers and sets seed only once.

Perianth: the petals and sepals. Both may be present, in which case there are said to be two perianth whorls, or there may be one. Which it is called is often a matter of choice, but in general green or brownish whorls are called sepals, and brightly coloured whorls petals.

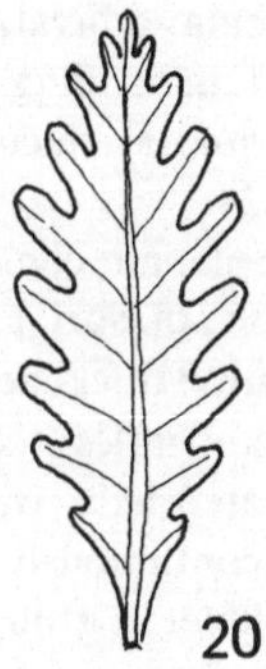

20

Perigynous zone: the part of the receptacle of a flower between base of ovary and inserti on of petals. Fig. 2.27.

Persistent: not caducous.

Petal: brightly coloured, usu. inner, perianth segment. Fig. 2:21,26,27,29.

Pinna, -ae: a leaflet of a pinnate leaf. Fig. 1:10.

Pinnate: (of a leaf), with the leaflets arranged like the branches of a feather. Fig. 1:10,11.

Pinnatifid: with lobes arranged like a feather, but not divided to the midrib. Text fig. 20.

Pinnatisect: like pinnatifid, but with some of the divisions to the midrib; part of the leaf pinnate, part pinnatifid. Text fig. 21.

Pollinia: gobular masses of aggregated pollen (in orchids).

Procumbent: lying down along the surface of the ground.

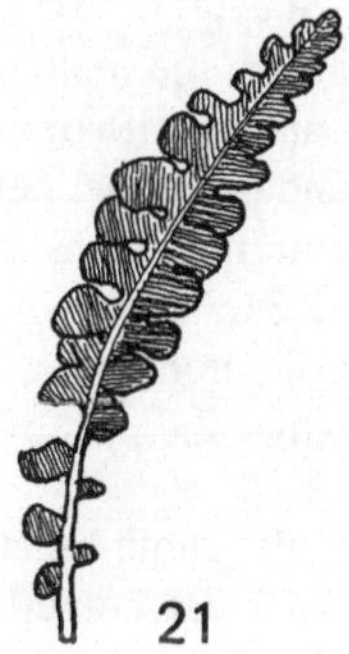

21

Pubescent: softly covered with short downy hairs.

Punctuate, dim. punctulate: dotted with small pits or protuberances.

Raceme: an unbranched inflorescence with pedicellate flowers, arranged so that the lowest ones come out first, newer flowers being produced at the tip (as a foxglove). Fig. 2:32.

Rhachis: the central axis of a compound leaf, Fig. 1:12, or of a spikelet.

Rhizome: an underground root-like perennial stem.

Sabre-shaped: text fig. 22.

Sagittate: arrow-shaped. Fig. 1:6.

Scarious: stiff and scale-like, not green.

Scorpioidal: coiled (like a scorpion's tail) usually of a cyme with the flowers arranged on either side of a coiled stem). Fig. 2:36.

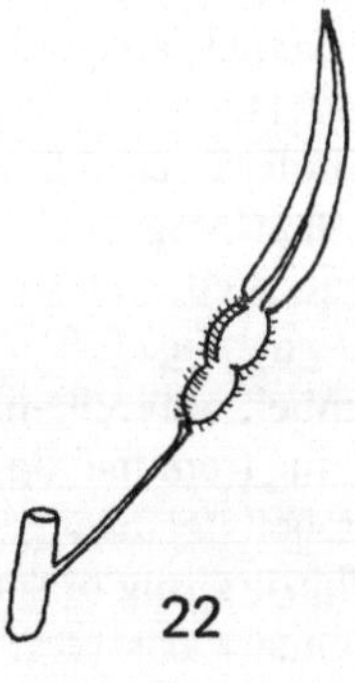

22

Secund: all pointing towards the same side (like the flowers on a bluebell).

Sepal: dull coloured, often outer, perianth segment, part of calyx. Fig. 2:21,26,27,28.

Serrate: saw-toothed. Fig. 1:5.

Serrulate: finely saw-toothed.

Shrub: a plant with several branched woody stems from at or near ground level.

Silicula: a bivalved pod less than 3 x as long as wide, with a septum dividing it internally into two compartments. Text fig. 23.

Siliqua: as silicula, but more than 3 x as long as wide. Text fig. 24.

Simple: not compound (of a leaf). Fig. 1:1-9.

Sorus, pl. sori: a group of sporangia. Text fig. 25.

Spadix: thickened axis of spike (Araceae). Text fig. 26.

Spike: inflorescence with sessile flowers arranged on an elongated axis, with the older flowers below and younger flowers continuously produced at the growing tip. Fig. 2:31.

Spikelet: a group of small florets arranged together on a common axis, esp. in Gramineae. Text fig. 27.

Sporangium: spore-producing bodies of Pteridophyta (including the fruiting bodies on under surface of a fern blade).

Spur: sac-like prolongation at base of flower (either from petal or calyx). Text fig. 28.

Stamen: the male reproductive part of a flower, comprising of an anther with or without a stalk (filament). Fig. 2:25.

Standard: upper petal of pea flower. See text fig. 19.

Stem: main axis of a plant, bearing leaves and flowers. Fig. 2:31,32,39.

Sterile: not fertile.

Stigma: the part of the gynoecium which collects the pollen. Fig. 2:28.

Stipule: small bract, often resembling a reduced leaf, at the base of a leaf, sometimes partly fused to the stalk. Fig. 1:11.

Style: the stalk of a stigma. Fig. 2:28.

Superior ovary: an ovary lying above the point of attachment of the perianth. Fig. 2:26.

Tendril: specialized part of plant, often derived from a leaf or leaflet, adapted for clinging on to surrounding plants etc. Fig. 1:11.

Trifoliate: used here to mean with three leaflets (of a compound leaf). Fig. 1:14,15.

Truncated: cut straight across, not indented or shaped in any way. Fig. 1:7.

Umbel: umbrella-like inflorescence (with a whorl of spokes arising from the top of the stem). Fig. 2:33.

Umbellate: arranged in an umbel. Fig. 2:33.

Whorl: a ring of similar structures all arising from a circle on the main axis (e.g. perianth whorl, whorl of stem lvs in Anemone, whorl of flowers in Labiatae etc) Fig. 2:39.

Wing: either a thin extension of the outline of a part, or the lateral, spreading petals of Orchidaceae, Leguminosae (see text fig. 19), Fumaria etc.

Valve: the separate flaps by which a fruit opens. Text fig. 20.

Zygomorphic: with two halves, one a mirror-image of the other. Fig. 2:23,24.

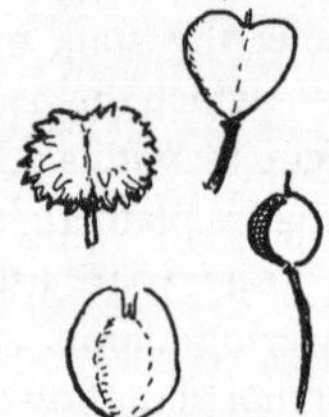
23

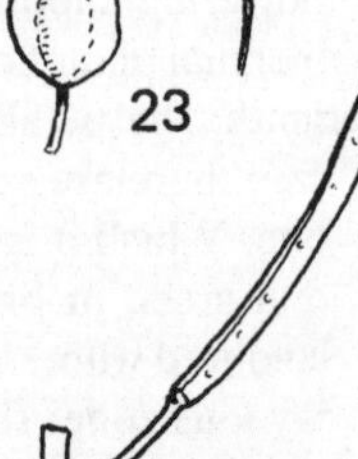
24

25

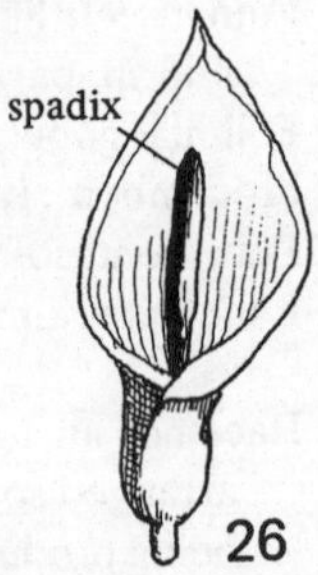

26

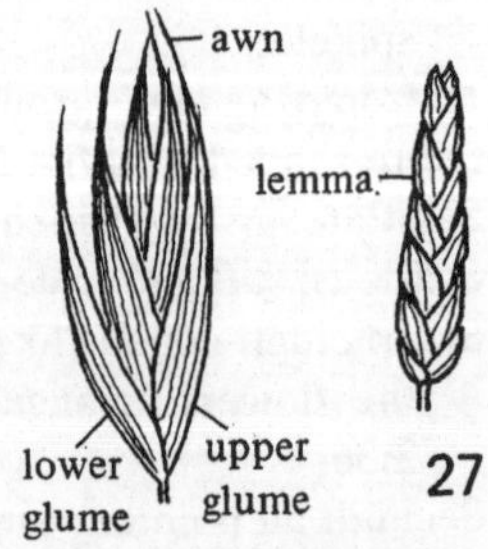

27

28

Bibliography

Barceló, F. 1879-81. *Flora de las Islas Baleares*. Palma de Mallorca.

Bonafè, F. 1977-80. *Flora de Mallorca*. Palma de Mallorca: Editorial Moll.

Bonner, A. 1982. *Plants of the Balearic Islands*. Palma de Mallorca: Editorial Moll.

Clapham, A.R. et al. 1962 (second ed.). *Flora of the British Isles*. Cambridge: University Press.

Grigson, G. 1958. *The Englishman's Flora*. Frogmore: Paladin.

Hubbard, C.E. 1954. *Grasses*. London: Penguin Books.

Hutchinson, J. 1959 (second ed.). *Families of Flowering Plants*. Oxford: Clarendon Press.

Knoche, H. 1921-23. *Flora Balearica*. Montpellier.

Polunin, O. 1969. *Flowers of Europe*. London: Oxford University Press.

Polunin, O. and B.E.Smythies 1973. *Flowers of South-West Europe*. London: Oxford University Press.

Schmeil, O. and Fitschen, J. 1959. *Flora von Deutschland*. Heidelberg: Quelle & Meyer.

Tutin, T.G. et al. 1964-80. *Flora Europaea*. Cambridge: University Press.

Local names: Most of these are from Bonafè (1977-80).

Others are from: Schmeil and Fitschen (1959), Grigson (1958), Polunin (1969) and Polunin and Smythies (1973).

Endemic species and subspecies

Aetheorhiza bulbosa ssp *willkommii* (Ma, I) 144, pl. 4.38
Allium grosii (I) 162
Aristolochia bianorii (Ma) 35
Asperula aristata ssp *pauii* (I) 112
Astragalus balearicus (Ma, Mi) 71, pl. 5.7
Avenula crassifolia (I) 183
Brassica balearica (Ma) 57, 61
Bupleurum barceloi (Ma, I) 104, pl. 2.27
Carex rorulenta (Ma) 195
Centaurea balearica (?Ma, Mi) 150
Chaenorrhinum rubrifolium ssp *formenterae* (Ma, I) 129
Crepis triasii (Ma, Mi) 153
Crocus cambessedesii (Ma, Mi) 164
Cymbalaria aequitriloba ssp *fragilis* (Mi) 130
Daphne rodriguezii (Mi) 92
Digitalis dubia (Ma, Mi) 128, pl. 5.46
Erodium reichardii (Ma, Mi, ?also in Corsica) 82, pl. 1.17
Euphorbia maresii (Ma, Mi) 86
Galium balearicum (Ma) 113
G. crespianum (Ma, I) 113
G. dorycnifolia (I) 71
Genista acanthoclada ssp '*balearica*' (Ma) 71
G. lucida (Ma) 71, pl. 5.4
Globularia cambessedesii (Ma) 131
Helichrysum ambiguum (Ma, Mi) 146, pl. 4.13
Helleborus lividus (Ma) 48, pl. 1.9
Hippocrepis balearica (Ma, Mi, I) 80, pl. 7.45
Hypericum balearicum (Ma, Mi, I) 93, pl. 2.18
Launaea cervicornis (Ma, Mi) 144, pl. 4.37
Ligusticum lucidum ssp *huteri* (Ma) 101
Limonium biflorum (Ma) 108
L. caprariense ssp *caprariense* (Ma) 108
L. caprariense ssp *multiflorum* (Ma, Mi, I) 108
L. inarimense ssp *ebusitanum* (Ma, I) 108
L. majoricum (Ma) 108
L. minutiflorum ssp *balearicum* (Ma) 108
L. oleifolium ssp *pseudodictyocladum* (Ma) 108
Linaria aeruginea ssp *pruinosa* (Ma) 129
Lonicera pyrenaica ssp *majoricensis* (Ma) 135
Lotus tetraphyllus (Ma, Mi) 78
Lysimachia minoricensis (Mi) 106
Paeonia cambessedesii (Ma, Mi) 51
Pastinaca lucida (Ma, Mi) 101
Phlomis italica (Ma, Mi) 121, pl. 5.31
Pimpinella bicknellii (Ma) 104
Plantago coronopus ssp *purpurascens* (Ma) 133, 135
Primula vulgaris ssp *balearica* (Ma) 106
Ranunculus weyleri (Ma) 50
Rhamnus ludovici-salvatoris (Ma, Mi, I) 90
Rubia peregrina ssp *angustifolia* (Ma, Mi, I) 112
Scutellaria balearica (Ma) 121, pl. 5.30
Senecio leucanthemifolius ssp *rodriguezii* (Ma, Mi) 148, pl. 4.22
Sibthorpia africana (Ma, Mi, I) 128, pl. 2.40
Teucrium asiaticum (Ma, Mi) 123, pl. 5.26
T. cossonii (Ma) 122, pl. 5.28
T. subspinosum (Ma) 122, pl. 5.27
Thymelaea myrtifolia (Ma, Mi) 92, pl. 3.17
Vicia bifoliata (Ma, Mi) 72
Viola jaubertiana (Ma) 93

Plants here with very limited distribution elsewhere:

Acer granatense (small area of Spain and Ma) 89
Arenaria balearica (Ma and Corsica) 44, pl. 2.2
Buxus balearica (small area of Spain and Ma) 90
Cephalaria squamiflora ssp *balearica* (Ma, Corsica, Sardinia) 137, pl. 4.5
Cyclamen balearicum (small area of France, Ma, Mi, I) 106, pl. 2.28
Naufraga balearica (Corsica and Ma) 100, pl. 2.25
Silene hifacensis (small areas of Spain and I) 46
S. pseudatocion (Ma, N.Africa and ?Spain) 47
Soleirolia soleirollii (Ma, Sardinia and Corsica) 33

Many plants here show features distinguishing them from populations elsewhere; some botanists tend more than others to separate some of these as distinct species and subspecies.

Index to keys

For genera with more than one species, the key to the genus is the last page number given.